AF770806

# TRACKSIDE around ST. LOUIS 1952-1959 with Jim Ozment

by James Sandrin

*The Market Street side of St. Louis Union Station looking to the southwest. Eighteenth Street is in the foreground. May 19, 1956.*

**Copyright © 2000**
Morning Sun Books, Inc.
**All rights reserved.** This book may not be reproduced in part or in whole without written permission from the publisher, except in the case of brief quotations or reproductions of the cover for the purposes of review.

MORNING SUN BOOKS INC.
ROBERT J. YANOSEY, President
To access our full library *In Color* visit us at
**www.morningsunbooks.com**

Published by
**Morning Sun Books, Inc.**
9 Pheasant Lane
Scotch Plains, NJ 07076

Library of Congress
Catalog Card No. 00-132092

First Printing
ISBN 1-58248-039-7

Color separation and printing by
**The Kutztown Publishing Co., Inc.**
Kutztown, Pennsylvania

THE KUTZTOWN PUBLISHING CO., INC.

**DEDICATION**

My dedication is to Joseph W. Sandrin and his uncles, who played baseball for the Blue Seal Coal Mine team in Livingston, Illinois. The rural farming and mining culture that spawned this rail network was the environment of my dad and his family. I hope this work recalls some fond memories on the eve of a new century.

**ACKNOWLEDGMENTS**

The intent to compile a book covering the railroad scene in a hub like Saint Louis, Missouri can only meet limited success, even when considering a limited window of time. When Bob Yanosey proposed this project, I had concerns about approaching the subject and the thoroughness of Jim Ozment's photo coverage of forty years ago. By definition though, by limiting the content to Jim's work, the book would meet its intended purpose. As it turned out, while the photography of J. L. Ozment is showcased, this book is a composite effort of some very knowledgeable and helpful people.

In addition to Jim's recollections, friends who provided suggestions and information include Edward DeRouin, Ed McLaughlin, James Singer and George Speir. The library of references used for background includes employee magazines and timetables, various issues of *Official Guides*, and most importantly *The Train Watcher's Guide to St. Louis* by John Szwajkart and the newsletters of the *Terminal Railroad Association of St. Louis Historical and Technical Society, Inc.* For those interested in the St. Louis railroad scene I recommend these two sources as starting points. As is usually the case, *Extra 2200 South* was used to check diesel locomotive rosters. Of the numerous books scanned for information during this project, one I recommend if you have the slightest interest in traction and the rural Midwest is *Illinois Terminal, The Electric Years* by Paul H. Stringham, published by Interurban Press. Other books which provided information that helped piece this puzzle together include Kalmbach's original *Diesel Spotter's Guide* and Morning Sun's *New York Central Trackside* with Eugene Van Dusen and *Illinois Central in Color* by Lloyd E. Stagner.

The format for this volume owes a debt to an engineering study sponsored by the TRRA in 1921. Titled *Report of Engineering Committee, St. Louis-East St. Louis Railroad Terminals* and published in 1922 by the Consulting Engineering firm of C. E. Smith and Company, this work provided concise information on all area trunk lines with maps detailing their operations within the operating sphere of the TRRA. Recommendations of the consulting firm for alterations to yards, bridge approaches, junctions and the like were all presented to the TRRA. For example, the River Front High Line and the southern approaches to the MacArthur Bridge, then known as Municipal Bridge, were study proposals acted upon by TRRA and consequently depicted in this book. Some of the maps presented in this volume were prepared as part of this study and are used by permission of Larry Thomas, editor for the TRRA Historical and Technical Society.

As is sometimes the case, errors and omissions will occur. These are my responsibility, for which I apologize.

James Sandrin
Milner, Colorado
December 1999

# TABLE OF CONTENTS

# TRACKSIDE *around* ST. LOUIS 1952-1959 *with* Jim Ozment

**ST. LOUIS GATEWAY**

Geography is both a benediction and a curse for the railroad center of Saint Louis, Missouri. Named for the canonized King Louis IX of France, St. Louis was founded in 1764 by Pierre La Clede Liguast at the confluence of the Mississippi and Missouri Rivers in what was then the Louisiana Territory. Its location guaranteed rapid growth as the Missouri River became the highway of commerce for the opening of the west. Appropriate to its role as a staging area for exploration, the "Gateway" claim was cemented after the Lewis & Clark Expedition embarked from and returned to the Illinois side of the Mississippi, opposite St. Louis. The early wagon trains which headed west along the Santa Fe, the Oregon and California Trails, all started at St. Louis, Missouri Territory. Oddly, when Missouri entered the Union, a more centrally located town, Jefferson City, was chosen as the capitol.

Saint Louis grew and prospered through the river boat years from 1817 through 1870. As technology progressed, the first railroads built into the bottoms region on the east bank of the Mississippi near what became East St. Louis in the 1850s. The first line on the western side of the river, a predecessor of the Missouri Pacific, opened for traffic in 1852. Railroads building from the northeast all vied for river front access in East St. Louis, resulting in a collection of small, stub ended yards lined up parallel to each other. These yards funneled traffic through the trackage of the Wiggins Ferry Company, which controlled the river front trackage and the car ferries that transferred railcars between the two river banks.

In addition to needing bridge access from the east, a major stumbling block to the development of St. Louis as a rail center was its location in a state sympathetic to the South. Northern financial interests were uneasy with the prospects of proposed railroad properties being within striking distance of hostile militia if war should break out with the Southern States. Consequently, as political tensions grew during the 1850s, new railroad construction pushed toward a Chicago gateway rather than St. Louis, insuring Chicago's spot as the rail hub of the nation. When hostilities did erupt, the St. Louis region proved to be out of harm's way although the city was kept under martial law by Union forces for the duration of the war.

**BRIDGES EVERYWHERE**

To overcome the limitations of the ferries, which were slow and subject to seasonal delays from flooding and river ice, the city had to wait for the person of Captain James B. Eads. A riverboat operator of genius and energy, his efforts were rewarded with the opening of the bridge bearing his name on July 4, 1874, only the second such structure on the entire river. Located at a position where it would least impact the river boat interests, the bridge piers kept the channel clear as best as could be accomplished, dividing the wharf area nearly in half for docking above and below the bridge adjacent to Washington Avenue. This location was not good from a railway engineering perspective since its west end rested upon rising ground that required a tunnel to reach the established rail yards in the Mill Creek Valley. After 1889 and the creation of the TRRA, the Eads Bridge came into its own as the TRRA used it for transfers between the Mill Creek Valley and Clearing Yard, north of Relay Junction.

It is quite remarkable that railroad executives of 1889 should have the insight to form a corporation such as this to handle interline transfers. That being the era of the so-called robber barons, it is historically unique that those men were able to set aside their egos for the collective good. The formation of the Terminal Railroad Association and the construction of the bridges and their connections are the elements that made St. Louis' rail operations unique and functional.

Even though the TRRA was in its formative years, concerns about the lack of transportation options in the St. Louis area caused the Merchants Exchange group to form the Merchants Bridge Terminal Railway Company in 1887. While the Merchants Bridge itself was completed between North St. Louis and Madison in May 1890, it did not prove to be a valid alternative to the Eads Bridge until early 1893 when the approaches were finished. Joined to trunk lines by Merchants Bridge Terminal Belt lines on both sides of the river, notable improvement in traffic flow occurred as cooperation with the other elements of the TRRA system developed. Interestingly, the Wiggins Ferry company, though now as a part of the TRRA, continued car float operations well into the 1900s despite the routing options the bridges provided.

To the north, the city of Alton desired to decrease its dependence on traffic sent through St. Louis. A bridge was built west of Alton, with aspirations for a westward outlet for locally generated business. Completed in late 1893, the bridge could not be used until connections with the Burlington at West Alton were finished in early 1894. The original company that finished the Alton Bridge was the St. Clair, Madison and St. Louis Belt Railroad, but subsequent financial problems resulted in the transfer of properties to the Missouri & Illinois Bridge & Belt Railroad, a consortium made up of some of the same roads which controlled the TRRA. This company became a subsidiary of the Burlington Route in the 1940s.

In 1910 the McKinley Bridge was completed to serve the lines of the Illinois Traction System, later to be renamed the Illinois Terminal Railroad. Connecting Venice, Illinois with North St. Louis, this bridge is south of the Merchants Bridge. The southern-most of the bridges over the Mississippi River is the Municipal Bridge, owned by the city of St. Louis. This double deck bridge was opened for traffic in 1916. The upper deck was for vehicular traffic , the lower level for rail traffic. After the outbreak of the Second World War, the structure was renamed the Douglas MacArthur Bridge. Its location is due east of the Mill Creek Valley, essentially a continuation of the alignment east of Union Station to Valley Junction on the Illinois side.

In review, the sequence of rail bridges from the north are: the Alton Bridge connecting West Alton, Missouri to Alton, Illinois; the TRRA Merchants Bridge connecting North St. Louis, Missouri to Madison, Illinois; the Illinois Terminal's McKinley Bridge connecting North St. Louis, Missouri to Venice, Illinois; TRRA Eads Bridge connecting St. Louis, Missouri to East St. Louis, Illinois via Relay Junction and finally, the City of St. Louis owned MacArthur Bridge connecting the Mill Creek Valley corridor in St. Louis, Missouri to East St. Louis, Illinois via Valley Junction.

**RAILROAD CO-OPERATION**

As a "Gateway" for both freight and passenger service, another element was needed to overcome the congestion caused by St. Louis' geography. In combination with the bridges, fluid operations are dependent on the terminal switching roads operating on both sides of the river. The previously mentioned Terminal Railroad Association is a consortium made up of the following trunk lines: Baltimore & Ohio - Southwestern; Chicago, Burlington & Quincy; Chicago, Rock Island & Pacific; Cleveland, Cincinnati, Chicago & St. Louis (Big Four-NYC); Gulf, Mobile & Ohio; Illinois Central; Louisville & Nashville; Missouri-Kansas-Texas; Missouri Pacific; Pennsylvania; St. Louis-San Francisco; St. Louis Southwestern; Southern; and the Wabash. Again, it was remarkable that the managers of these railroads could come to such an agreement and create the TRRA for the benefit of all the St. Louis region, putting vested interests aside. While the agreements between the various roads differed with types of traffic, the configuration of yards, interchanges and various work rules governing crew assignments, the TRRA and Alton & Southern kept St. Louis railroads fluid and greatly reduced the congestion born of individual roads doing their own transfer movements via the bridges and the many individual yards.

The showpiece of the TRRA was St. Louis Union Station *(page 1)*. Built in 1893, St. Louis Union Station is a beautiful structure of cut grey granite, facing Market Street to the north and adjoining Eighteenth Street on the east and Twentieth Street on the west. Numerous suggestions were made through the years to build outlying depots to reduce traffic into this central location, but none were ever acted upon, although Union Station was enlarged in 1903 prior to the Louisiana Purchase Exposition, the centennial celebration of the Louisiana Purchase. Dimensions resulting from those modifications were 32 tracks arranged in pairs between platforms. Stub-ended, each track is 775-feet long within the train shed, half extending to 850-feet in total length, the other half extending to 900-feet in total length. All tracks are accessed by two interlaced, multiple track wyes. Baggage and express facilities were built on the east side of the depot, with separate trackage served by the Terminal Railroad. All trains using Union Station stopped, were wyed, and then backed into the train shed. While this seemed an operating inconvenience for arrivals, this practice kept locomotives away from the passenger platforms and allowed movement of motive power independent of their trains. Head end switching and the spotting of consists was done by the Terminal Railroad switch engines.

**JIM OZMENT - PHOTOGRAPHER**

Due to the special interest of his paternal grandmother, Jim Ozment became a seasoned streetcar rider at an early age. Since the St. Louis of the 1950s still had extensive traction operations, Jim sought out and photographed the trolleys and interurbans at every opportunity. Running on streets, segregated rights of way or interchanging with steam roads, the St. Louis Public Service, the St. Louis Water Works Railway and the Illinois Terminal Railroad provided ample operations under wire to attract and seduce photographers during their final years of operation.

A brief glimpse of the early years of the Museum of Transport is included. Jim and his friends were regular volunteers during the early 1950s and many of them are still active there today.

We try to present these photographs and the supporting text in the present tense. It would be quite difficult to provide a complete historic accounting of the roads shown in the years subsequent to these photos. Coverage represents the interests and opportunities available to Jim during the transition years of steam to diesel. Limitations included the time and accessibility of certain facilities and parts of town available to a college student. In 1956 Jim began working summers for the Denver & Rio Grande Western until his graduation from Washington University in 1958. His adventures out west and more biographical data were highlighted in Morning Sun's *Rio Grande Trackside with Jim Ozment (#9)*.

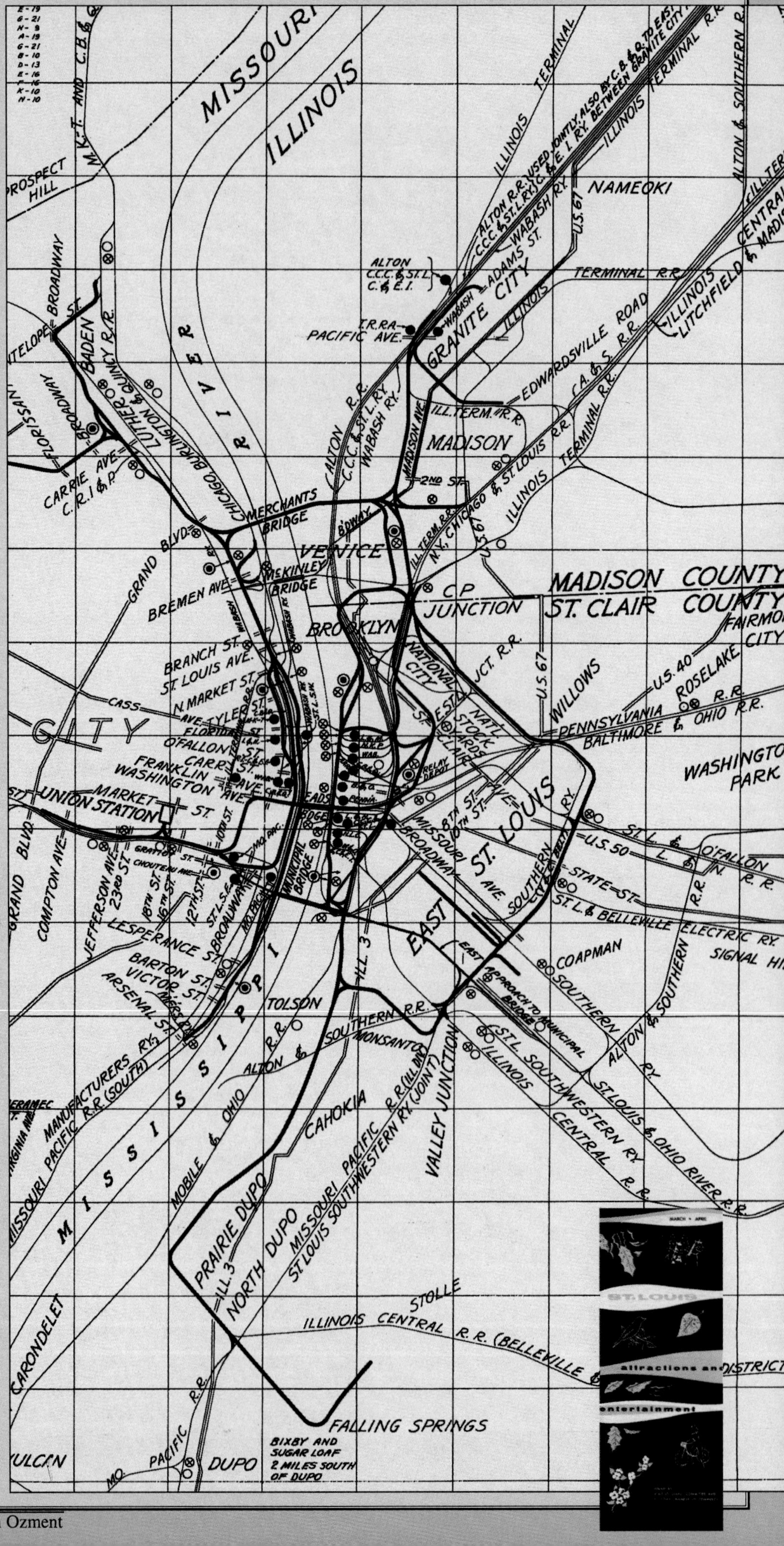

On October 3, 1954 a string of meat reefers dominate the view to the north of the Mill Creek Valley at Grand Avenue. This skyline is west of the downtown area. The Mill Creek water course provided a route for the first railroad construction on the west side of the Mississippi at St. Louis.

The first bridge in the region, Eads Bridge had to contend with a less than ideal location to accommodate the river boat traffic on the west bank and congested yard arrangements on the east side. Streetcar tracks have recently been removed from the upper deck in this view towards East St. Louis taken September 19, 1954. The last previous owners of those streetcar tracks operated in 1935. Curving to the left, Eads' lower deck trackage descends toward Relay Junction to the northeast. To the right, Louisville & Nashville steam locomotives are parked in dead lines at Lower Yard, the riverfront connection with the Wiggins Ferry trackage of the TRRA.

**V**iewed from the rear of the eastbound Wabash BANNER BLUE, Merchants Bridge quickly recedes into the west. This was recorded from rear of the parlor/observation car *City of Wabash* entering Illinois on October 1, 1959.

**L**ooking south along the St. Louis wharf, the American Royal Shows Circus train heads west on the lower deck of the MacArthur Bridge on June 4, 1956. A TRRA switcher provides the power for this special movement. The MacArthur Bridge is aligned with the Mill Creek Valley corridor and is the most direct route to East St. Louis.

The American Royal Shows Circus train rolls off the MacArthur Bridge onto the southern high line approach bridges. The approaches were built as a result of the 1921 Engineering Study undertaken by the TRRA to ease congestion along the Mississippi River front. These bridges bring the trackage down to the Missouri Pacific line into Lesperance Yard, south of the downtown area along the Mississippi River.

# Alton & Southern Railroad

ST. LOUIS • EAST ST. LOUIS
Alton and Southern
GATEWAY

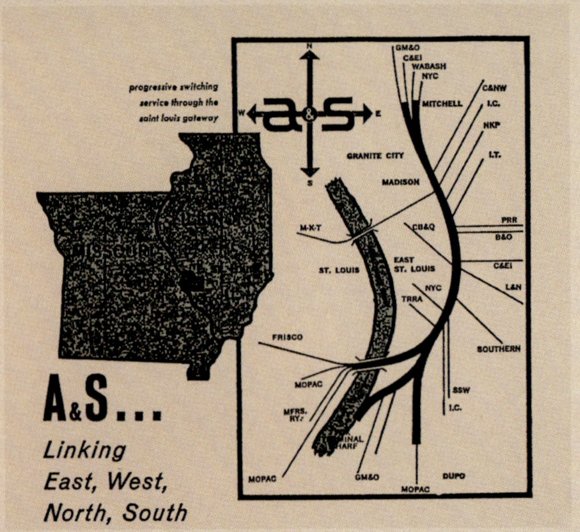


The Alton & Southern is an outer belt terminal switching line serving East St. Louis and the roads that radiate from it. Starting on the south side of town, the line begins at the Fox Terminal wharf on the Mississippi. Formerly a car ferry pier, this facility serves as a grain loading terminal for barges.

Crossing the GM&O at Cahokia, the line runs northeast via Valley Junction to Davis Yard, the A&S main yard and engine terminal. Parallel to and north of the Illinois Central and Cotton Belt yards, Davis sits at the foot of the eastern approach to the MacArthur Bridge.

Leaving Davis Yard, the line turns north, crossing (in order): the Southern, B&O, Pennsy, Nickel Plate, IC, IT, L&M (C&NW), and several industrial roads. The northern terminal for the road is the interchange yard at Mitchell, Illinois. There the A&S receives from or delivers to GM&O, CB&Q, C&EI, CCC&StL (NYC), IT and Wabash. In addition to interchange with these trunk lines, the A&S serves many online customers directly and delivers to the TRRA. All this work is performed by a roster of Alco-GE RS-2 and RS-3 road switchers. The RS-2s were numbered 28-42, and 46-54. Three RS-3s on the roster were numbered 43-45.

A&S RS-2 31 leads a transfer across the diamonds at Mitchell, Illinois on May 4, 1958. Northeast of Granite City, Mitchell is the northernmost station on the A&S. Traffic from the GM&O, CB&Q, C&EI, CCC&StL, IT and Wabash all funnel through this point to Madison and East St. Louis. Leaving the nose access doors open, the crew no doubt hopes to improve cab ventilation on this typically humid, spring day.

Alco-GE RS-2 41 leads 28 cars south from Mitchell on January 3, 1956.

This scene defines the A&S. On May 19, 1956 units 39 and 34 delivered interchange traffic to the TRRA at Mill Creek Valley in St. Louis. A&S interchange in St. Louis was accomplished by trackage rights over the TRRA and the MacArthur Bridge. Trackage owned or controlled by the A&S was limited to the Illinois side of the Mississippi. With the interchange of traffic, these two terminal roads give credence to Saint Louis' "Gateway City" claim, serving as intermediaries for the numerous trunk lines.

# BALTIMORE & OHIO

The Baltimore and Ohio enters the East St. Louis area from the east side, south of and parallel to the Pennsylvania Railroad. Technically the B&O-SW, or Southwestern District, this line originates in Cincinnati, Ohio and runs due west across southern Indiana and Illinois. At Shattuc, Illinois the line crosses the Centralia line of the Chicago, Burlington & Quincy. West of Shattuc, the Burlington had trackage rights over the line to Relay Junction in East St. Louis for movement of coal from the southern Illinois coal fields.

Cone Yard, east of Relay Junction, was the primary yard and engine terminal for the B&O in East St. Louis. A smaller yard, located north of Eads Bridge, connected with the TRRA Wiggins Ferry Yard on the river front.

All passenger power operated into Union Station and then returned light to Cone Yard for servicing. Named trains operating into St. Louis included: THE NATIONAL LIMITED, THE DIPLOMAT, and simply THE MAIL TRAIN.

Shot from the Fourteenth Street Viaduct, a pair of E-8 passenger units coupler covers wide-open head east light after bringing their train into Union Station on September 2, 1954. They will be fueled, sanded and inspected at Cone Yard.

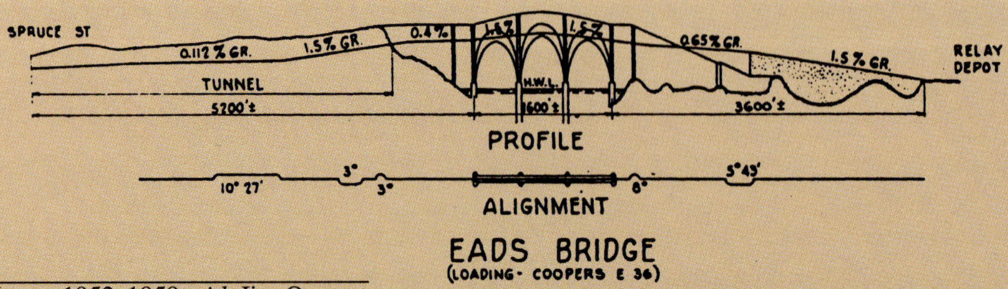

**S**everal eastern roads assigned steam locomotives to the St. Louis area as their retirement drew near. This made optimal use of the remaining engine facilities: coal towers, ash pits, round houses and the like. It also made the final transfer to Granite City Steel easier when scrapping occurred. On March 6, 1955 2-8-0 2818 quietly steams at East St. Louis while assigned to yard duty.

**A**fter servicing at Cone Yard, Pacific 5222 waits to follow a Pennsy mail train west over Eads Bridge on May 16, 1956. The Pacific will take train No. 29 east later that day.

## NATIONAL LIMITED

**Pullman and Coach—Dieseliner**

Duplex-Roomettes, Roomettes, Bedrooms, Compartments, Drawing Rooms, Berths and Sections. Individual Reclining Coach-Seats; Observation-Lounge Car. Stewardess-Nurse.

| No. 1 WESTWARD (Read down) | STANDARD TIME | | No. 2 EASTWARD (Read up) |
|---|---|---|---|
| 5,20 p.m. | Lv. Baltimore (Mt. Royal Sta.) | Ar. | 7.45 a.m. |
| 5.25 p.m. | Lv. Baltimore (Camden Sta.) | Ar. | 7.40 a.m. |
| 6.30 p.m. | Lv. Washington | Ar. | 6.40 a.m. |
| 7.25 a.m. | Ar. Cincinnati | Lv. | 6.10 p.m. |
| 10.05 a.m. | Ar. Louisville (Central Sta.) | Lv. | 1.35 p.m. |
| 1.15 p.m. | Ar. St. Louis | Lv. | 10.05 a.m. |

## DIPLOMAT

**Pullman and Coach—Dieseliner**

Duplex-Roomettes, Bedrooms, Berths and Sections. Reclining-Seat Coaches, Lounge Car.

| No. 3 WESTWARD (Read down) | STANDARD TIME | | No. 4 EASTWARD (Read up) |
|---|---|---|---|
| 8.00 p.m. | Lv. Baltimore (Mt. Royal Sta.) | Ar. | 12.15 p.m. |
| 8.05 p.m. | Lv. Baltimore (Camden Sta.) | Ar. | 12.10 p.m. |
| 9.15 p.m. | Lv. Washington | Ar. | 11.10 a.m. |
| 9.45 a.m. | Ar. Cincinnati | Lv. | 9.55 p.m. |
| 12.00 noon | Ar. Louisville (Central Sta.) | Lv. | 5.35 p.m. |
| 3.45 p.m. | Ar. St. Louis | Lv. | 1.45 p.m. |

Immaculate, especially on this late date, Pacific 5219 sits at Cone Yard Engine terminal on May 24, 1956. Just beyond the tender, some EMD replacements wait for their next call to service.

Pacific 5222 lays down a splendid plume of coal smoke as it races east with train No. 29, two miles east of Fairmont City, Illinois. On May 16, 1956 train 29 was all mail and express with a single coach bringing up the rear.

Approaching the CB&Q crossing at Shattuc, GP7 921 rolls a local freight east on December 30, 1955.

The B&O depot and tower at Shattuc were well maintained when photographed on December 30, 1955. In addition to B&O traffic, clerks here also issued orders and clearances for Burlington coal traffic headed for East St. Louis. Those Q stock cars in the background are on the interchange track east of the crossing.

# Chicago & Eastern Illinois

C&EI

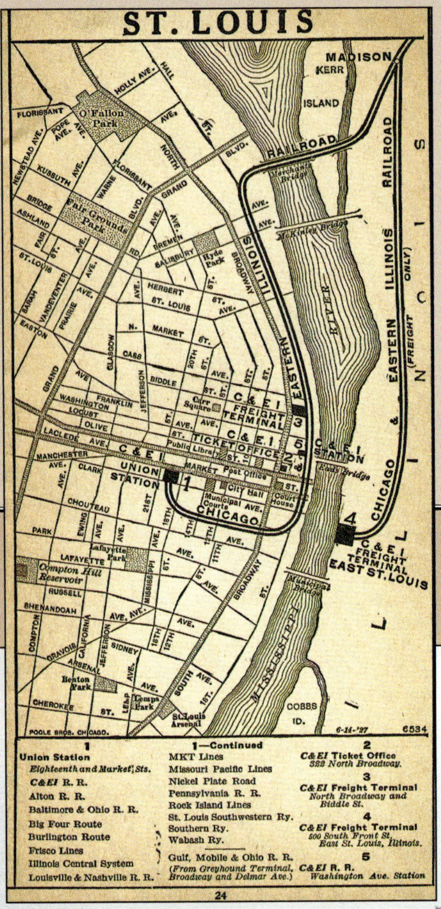


Self described as the "Boulevard of Steel," the Chicago & Eastern Illinois was a small bridge road that ran south from Chicago to Woodland, Illinois. At Woodland Junction, the double track mainline split into two main routes. One ran south to Evansville, Indiana to the Louisville & Nashville Railroad. The second line angled southwest across Illinois through Findlay to Pana, the junction with the CCC&StL. South of Pana the C&EI used trackage rights over the CCC&StL to Mitchell, five miles north of Granite City, Illinois. From Findlay the C&EI line went south through the Illinois coal fields to connect with the Mopac and Cotton Belt at the Mississippi River crossing of Thebes, Illinois.

At Mitchell the C&EI maintained a yard and engine terminal. Operating south of Granite City, the C&EI delivered its own interchange to the Frisco, rather than use the TRRA. This unique operation was a holdover from the time when both C&EI and Frisco were controlled by the same financial interests. Using C&EI locomotives, transfers were operated via the Merchants Bridge to Twenty-third Street in the Mill Creek Valley, the east end of the SL-SF (Frisco) Ewing Avenue Yard. Until 1949, the C&EI operated an overnight train between Chicago and St. Louis called THE ZIPPER. Never dieselized, THE ZIPPER was a heavyweight, mail and express train.

Looking east from the Jefferson Avenue Viaduct, C&EI 200, a rare EMD BL2 road switcher, works a short cut of cars for the Frisco at Ewing Avenue Yard on May 30, 1957. Left of the locomotive are the TRRA coach yards. This rail corridor is the Mill Creek Valley.

Southbound near Mitchell, Illinois, C&EI GP7 204 singlehandedly leads a lengthy freight on January 3, 1956. The Geep is running on CCC&StL (NYC) rails at this location.

South of Mitchell, a southbound freight rolls through the outskirts of Granite City on December 8, 1957. Caboose 325 carries the markers on that gray, early winter day.

# Chicago & North Western

Chicago & North Western operated into East St. Louis under the guise of the Litchfield & Madison, a small coal hauling road which the C&NW system would absorb in 1958. Traffic was delivered to the yard in Madison, south of Granite City. Westbound traffic moved via the TRRA over the Merchants Bridge. East St. Louis traffic moved via the A&S interchange adjacent the Madison yard.

Before the official takeover of the L&M in 1958, C&NW operated the L&M as part of the SI, the Southern Illinois district of the Galena Division. Daily freight trains 380 and 381 used C&NW crews between Nelson, Illinois, on the Chicago - Omaha mainline, and Benld. South of Benld, L&M crews manned these trains for the last 40-miles into Madison.

South of Edwardsville, the L&M ran parallel to the Nickel Plate to Glen Carbon. At Glen Carbon, the L&M joined the Illinois Central for a shared trackage operation into Madison.

**Table 25.**
**NELSON-ST. LOUIS.**

| Mls. | STATIONS. |
|---|---|
| 0 | **Nelson** |
| 6.9 | Van Petten |
| 11.7 | Hahnaman |
| 16.7 | Normandy |
| 24.0 | Manlius |
| 30.3 | Langley |
| 34.4 | Buda |
| 40.8 | Morse |
| 46.8 | Broadmoor |
| 51.4 | Camp Grove |
| 57.9 | Speer |
| 63.7 | Akron |
| 72.1 | Pioneer |
| 85.0 | **Peoria** |
| 83.1 | Keystone Switch (*Peoria*) |
| 86.0 | Sommer |
| 93.5 | South Pekin |
| 106.5 | Allen |
| 114.1 | Luther |
| 118.6 | Hubly |
| 124.2 | Sweetwater |
| 127.4 | Culver |
| 130.8 | Barr |
| 143.6 | Archer |
| 151.0 | Lick |
| 157.5 | Compro |
| 163.6 | Virden |
| 167.7 | Girard |
| 180.1 | Womac |
| 192.0 | **Benld** |
| 202.8 | **Staunton** |
| 208.0 | **Mt. Olive** |
| 202.7 | Williamson |
| 204.4 | **Livingston** |
| 203.7 | Worden |
| 208.9 | Manix |
| 215.1 | **Edwardsville** |
| 219.6 | Glen Carbon |
| 223.4 | Stallings |
| 226.9 | Granite City |
| 229.6 | **Madison** |
| 232.9 | **East St. Louis** |
| 236.0 | **St. Louis** |

Northbound train 380 passes through Edwardsville on a dismal February 19, 1956. Jim took this shot from the Illinois Terminal tower protecting the IT/L&M crossing. Up front on this dark morning are the 4069, 4068B and 4077C, a typical three-unit lashup assigned this train. On this day No. 380 has 121 cars.

On the west side of Horseshoe Lake, an isolated remnant of a Mississippi River meander predating U.S. Army Corps of Engineers river control efforts, northbound 380 accelerates two miles north of Granite City. L&M caboose #7 brings up the rear on this March 27, 1956 version of No. 380. This was the era of specific crew assignments for cabooses, so No. 7 would be switched out at Benld and replaced with a North Western waycar for the trip north. "Waycar" was the preferred term for caboose on both the North Western and Burlington.

Southbound train 381 approaches our location at Horseshoe Lake on March 27, 1956. Power for this train was the matched 4078C, B and A set. Included in both consists are piggyback loads immediately behind the locomotives.

# CHICAGO, BURLINGTON & QUINCY

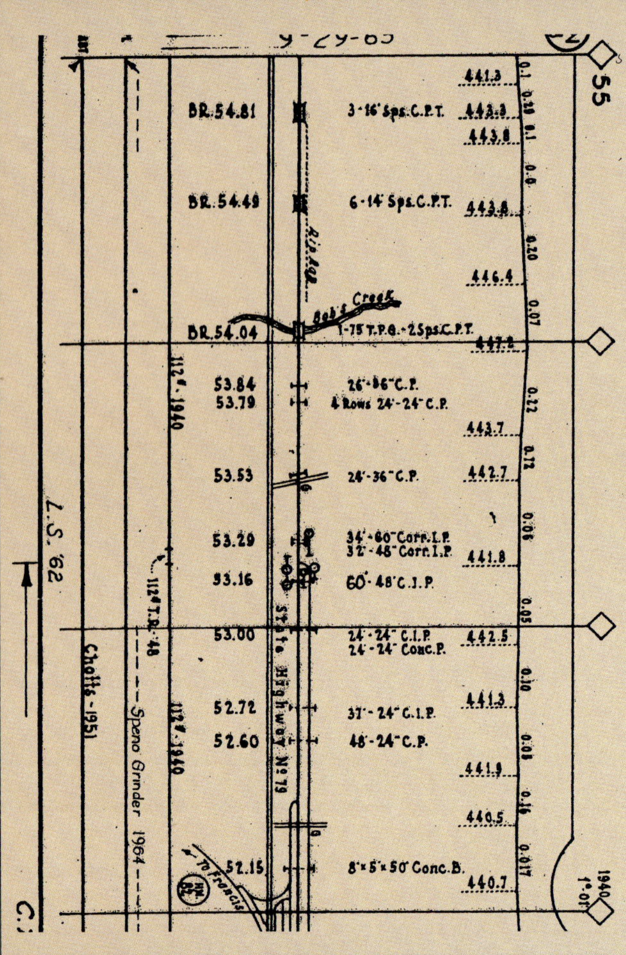


The CB&Q entered the St. Louis area over three routes from three different directions. Two of the lines served East St. Louis from Illinois. The northernmost of these lines and a Missouri line accessed north St. Louis and Union Station.

The CB&Q enters St. Louis proper over its river line from the north. Old Monroe, Missouri is the junction for the Kansas City - St. Louis line and the line that follows the Mississippi south from Hannibal, Missouri. This is a single track line. At Machens, Missouri the MKT line from St. Charles joins the Q for a shared trackage arrangement along the west bank of the river to West Alton. West Alton is the junction with the line that crosses the river via the Alton Bridge. That line is operated by the Missouri and Illinois Bridge & Belt Railway, a jointly owned subsidiary of the CB&Q and Missouri Pacific. About a mile south of West Alton, the line crosses the Missouri River on the Bellefontaine Bridge. Continuing southeast, the MKT leaves the Q for its own yard at Baden. At North St. Louis, north of the Merchants Bridge, the Q has its main yard and engine terminal. Traffic destined for the Mill Creek Valley or Union Station follows the TRRA High Line along the river from North St. Louis.

East St. Louis is served by two lines. The first line comes from the east, via trackage rights over the Baltimore & Ohio from Shattuc, Illinois. This line feeds directly into the Burlington's East St. Louis yard through Relay Junction. By this trackage rights arrangement, the Q avoided hauling St. Louis-bound coal from Centralia round-about through Beardstown.

A second line for Galesburg or Kansas City traffic, either uses the Merchants Bridge or the route through Bridge Junction and follows the GM&O (formerly the Chicago & Alton) line north to Alton, Illinois. This routing passes through East Alton and the Junction of the M&IB&B line from the Alton Bridge. A pairing of the GM&O line with the CCC&StL line through Edwardsville, creates a double track line north of Mitchell. Continuing north of East Alton, Burlington trains follow their own rails north to Concord Junction and Beardstown, Illinois.

Transfers from North St. Louis to East St. Louis were routed over the Alton Bridge. This forty-mile round trip took a full eight hours, but was still preferred to dealing with the congestion of transfers via the TRRA, in spite of the Q being one of the TRRA owner roads.

After receiving a green indication from the interlocking signal in the background, Q diesel-electric motorcar 9841 approaches the Shattuc depot on December 30, 1955. This car provided passenger, mail and express service for Centralia and the line south to Herrin, Illinois.

*Silver Charger*, CB&Q 9908, leads train No. 43 around the east leg of the wye at Union Station on May 19, 1956. Photographed from the Eighteenth Street Viaduct, No. 43 will follow the west bank of the Mississippi River north to Burlington, Iowa.

Burlington 9903, THE MARK TWAIN ZEPHYR, departs Union Station in 1952. This train covered the St. Louis - Burlington, Iowa run continuously for eighteen years, starting in 1935. The daily round trip covered 442-miles. *(Charles E. Able Photograph, JL Ozment Collection)*

**B**urlington train No. 43, powered by E7 9925B, passes a pair of Pennsy E8s on September 2, 1954. TRRA's Fourteenth Street Engine Terminal, about five blocks east of Union Station, was the site of this action. Most Burlington passenger power was serviced there except for Shovelnose units like the 9903. Since those units were permanently articulated with their trains, the entire consist was serviced at North Rankin Avenue, west of Union Station. The TRRA switcher in the background is switching the U. S. Mail facility on the east side of the depot.

**R**olling north on the TRRA High Line, THE MARK TWAIN ZEPHYR heads for home rails on January 30, 1958. This trackage served as the approach for the Merchants Bridge and avoided river front industrial trackage and grade crossings. Burlington's own line branched off this line south of the McKinley Bridge and passed under the west ends of both the McKinley and Merchants Bridges. The MTZ train set had only four more months of service before its retirement in April 1958.

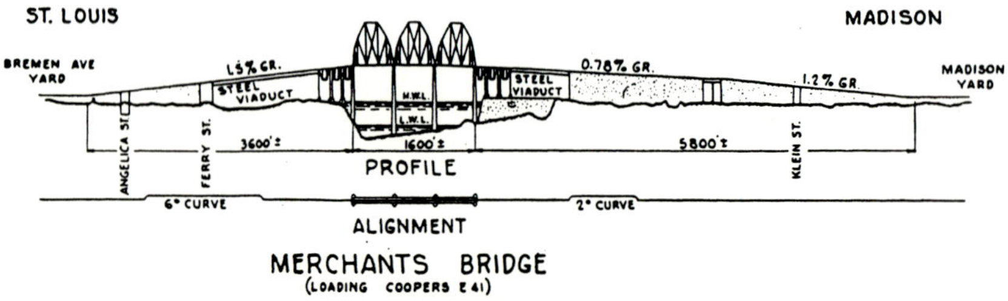

On the east side of the river, train No. 47 bound for Beardstown, Illinois rolls over the CCC&StL at Mitchell. On May 4, 1958 EMD E7 9928B powered this run. After leaving Union Station, this train followed the High Line and crossed over Merchants Bridge. Once on the east bank of the Mississippi, No. 47 followed the "Big Four" line north to East Alton.

Northbound for Galesburg on January 3, 1956 the 101 cars of train No. 92 follow F3 160 at Mitchell, Illinois. Waycar 13878 brings up the rear as the train passes the approach signals for Lenox Junction. Big Four trackage splits at Lenox Junction, the passenger cutoff going northeast to Hillsboro, while the original line goes north to East Alton. Burlington traffic followed the original line.

Also northbound at Mitchell on January 3, 1956 was passenger local No. 47. Instead of the usual E-unit, today's power is boiler-equipped GP7 222. Trailing the short consist is C&NW Pullman *Huron*. Until 1953, the bridge piers bracketing the joint line supported the Illinois Terminal electric interurban line to Alton.

West of Union Station in the Mill Creek Valley is TRRA's Rankin Avenue coach yard and engine terminal. Burlington's articulated train sets were serviced here. The equipment to the left is for the ZEPHYR-ROCKET, a joint Burlington - Rock Island operation between St. Louis and Minneapolis, Minnesota.

North St. Louis was the west side location for Burlington's yard and engine terminal. Still in its infancy, TOFC service was offered to customers and the necessary facilities were provided adjacent to the yard. On December 1, 1957 these trailers were ready for shipping at North St. Louis.

**D**aily operations required providing instructions to train crews to keep them informed as to changing work orders or the movements of opposing trains. Such an information transfer is performed once again at sundown on March 20, 1956. An operator at West Alton Tower delivers orders on the fly to the rear end crew of east bound No. 72. This train is the daily freight between Lincoln (Nebraska) and East St. Louis.

**O**n the river line north of St. Louis, the 9908 and train No. 43 stop at Louisiana, Missouri on June 4, 1957. Originally purchased for the GENERAL PERSHING ZEPHYR, a St. Louis - Kansas City train, the 9908 survived the downgrading of service to handle the short heavyweight consist, seen on a typical sultry summer day.

In central Missouri, the Bevier & Southern hauled coal with diminutive 2-6-0s, leased Q Mikados and Burlington rolling stock. In appearance the little road looked for all intents and purposes to be a Burlington branch line. At Bevier, the junction of the Q and B&S, a standard Burlington water tank still serves on April 2, 1958 as B&S 112 approaches from the west.

In a steam era ritual, the engineer does a walk-around inspection while the fireman fills the tank before heading out to the mine on April 2, 1958.

Later that same day, the 112 leads a train of loads north toward the interchange at Bevier.

On June 4, 1958 leased Burlington Mikado 4955 steams at the Bevier Mine near Binkley, Missouri. As loading progresses, the Mike will return to switching cars for the tipple crew and assemble them into a train for the run back to Bevier.

# Cleveland, Cincinnati, Chicago & St. Louis

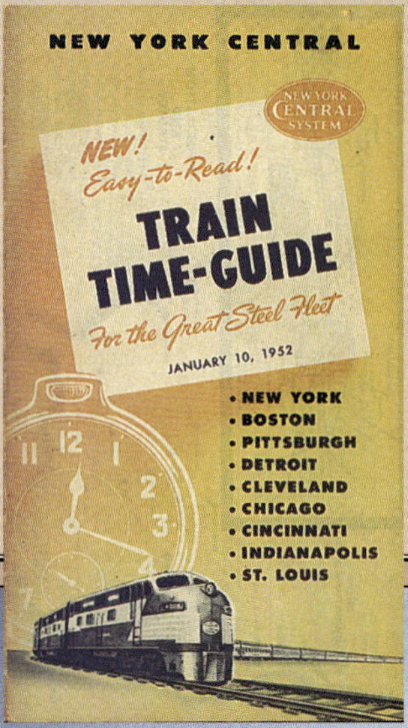


The Cleveland, Cincinnati, Chicago & St. Louis Railway, better known as the BIG FOUR, was the southwestern component of the New York Central system that served East St. Louis. Leaving Indianapolis on a southwestern heading, this line passes through Terre Haute, Paris, Mattoon (the Division and crew change point) and Pana to reach Hillsboro, Illinois. At Hillsboro, the original line ran west to Litchfield, Gillespie, East Alton and Wood River. The "new" cutoff went southwest from Hillsboro. Old and new lines came together again at Mitchell. South of Mitchell, the line passes through Granite City and Madison to reach East St. Louis via Bridge Junction. Eastbound traffic departed from a small yard just north of Bridge Junction. The main yard was south of Eads Bridge with access through Relay Junction and the river front Wiggins Ferry yard. Freight locomotives were serviced here while passenger power was serviced at Fourteenth Street by the TRRA.

Named New York Central passenger trains serving St. Louis included THE SOUTHWESTERN LIMITED, THE GATEWAY and THE KNICKERBOCKER.

On March 6, 1955, Jim maneuvered a ride on one of the East St. Louis switch runs. Seen from the cupola of caboose 17665, Class U-3 0-8-0 7621 was the motive power. As steam worked its final months on the Big Four, switchers were transferred west, to consolidate power where facilities and personnel were available to service them. As important, though certainly a darker aspect, this placed the locomotives nearer the scrap yards and steel mill which would be their final destination.

That same day, caboose 17665 poses at East St. Louis after the transfer work is finished.

Jim recorded these views of the 7621 working on March 6, 1955. One detail of the job, an everyday occurrence forty-five years ago, was the pin puller riding the footboards to the next coupling. It is easier to find active steam power in the 1990s than find a switch engine with footboards.

**A**nother Class U-3 assigned to East St. Louis was the 7799. Jim found the 0-8-0 working a cut of cars on December 30, 1954. Relay Depot is to the left beyond the bridge.

**D**owngraded road power also congregated to the west as retirement drew near. A Class H-5 Mikado, the 1571, works the yard leads near Relay Depot on December 30, 1954.

**O**n March 6, 1955 the 1571 was still working the East St. Louis yards around Relay Depot. Passengers departing from East St. Louis could make their connections there since traffic coming off Eads or Merchants bridges could stop at Relay. The name comes from the steam era practice of cutting off road power here and allowing TRRA switchers to take passenger trains through the Eads Tunnel and into Union Station, much like a relay race.

During the Fifties and early Sixties, New York Central passenger power was serviced at TRRA's Fourteenth Street Engine Terminal. The distant headlight belongs to Burlington train No. 47, departing Union Station on the eastbound passenger main. Class J-1 Hudson 5296 waits next to a pair of replacements, in this case an E8/E7 consist in the then current lightning stripe scheme, on a murky December 22, 1955. The Hudson was waiting for the call to power a holiday mail extra east.

South of St. Louis, at the southern extremity of Illinois at the confluence of the Mississippi and Ohio Rivers, is Cairo, Illinois. On June 18, 1958 a rare FM H20-44, number 7117, idles at the small engine terminal there. This location was literally the end of the line for the Big Four in Illinois.

Westbound at Mitchell on that same January 3, 1956, this freight was powered by a pair of Baldwin Sharks, the 3808 and 3815. These RF-16 cab units were members of a small fleet of eighteen such units, numbered 3804 - 3821, acquired in 1951 and early 1952. After a trial period on the Chicago - New York mainline, most of the Central's Baldwin diesel locomotives were transferred to the Big Four where tractive effort was more important than speed. A few of these veterans would survive until the late 1960s to be sold off to the Monongahela Railway in southwestern Pennsylvania.

**D**ecember 1, 1957 saw this westbound, heavy with head end business, follow E8 4093 off the approach for the Merchants Bridge. Immediately, the westbound rolls onto the River Front High Line for the final leg of the trip to Union Station.

Mitchell is the location for this view of the SOUTHWESTERN LIMITED on January 3, 1956. Instead of the usual E-units, today two steam boiler equipped GP9s, 5940 and 5944, power the eastbound.

# Gulf, Mobile & Ohio Railroad

GM&O was a union of two roads that came together at East St. Louis. Coming from the north, the Chicago & Alton, or simply "The Alton" entered the region through Alton, Illinois and Granite City. The southern member was the Gulf, Mobile and Ohio, a composite of the Mobile & Ohio and the Gulf, Mobile & Northern, entering the area on the south side of East St. Louis. The two combined gave the GM&O the only true north - south through line in the area.

From Roodhouse, Illinois, the junction with the C&A Springfield-Kansas City line, the northern line runs south to Godfrey, Illinois where it meets the passenger line from Springfield. South of Godfrey, the line runs through Alton to Wood River where it joins the Big Four's original mainline for a paired trackage arrangement into Mitchell and south to Venice, Illinois, the site of the main yard and engine terminal. The Brooklyn and Bridge Junction Yards were used for interchange via the TRRA. GM&O had no trackage or facilities on the west side of the river. Passenger traffic entered St. Louis via the Merchants Bridge and the High Line to Union Station.

The southern partner, the Gulf, Mobile and Ohio, paralleled the Mississippi from the junction at Jackson, Tennessee north to the Ohio River crossing at Cairo, Illinois. North of Cairo, the line ran roughly parallel to the river, though a few miles to the east over Alto Pass. Freight traffic was handled at three yards on the south side: southernmost was the Cahokia yard, near the town of the same name; then Wood Yard, south of Eads Bridge and Inside Yard on the river front. Trackage rights over the TRRA through Relay Junction allowed access to Union Station.

Passenger traffic into Union Station was brisk during the 1950s. Chicago-St. Louis traffic included THE MAIL, THE MIDNIGHT SPECIAL, THE ABRAHAM LINCOLN and THE (ALTON) LIMITED. South of St. Louis, THE GULF COAST REBEL traveled the route once served by the lightweight REBELS, built by ACF in 1935.

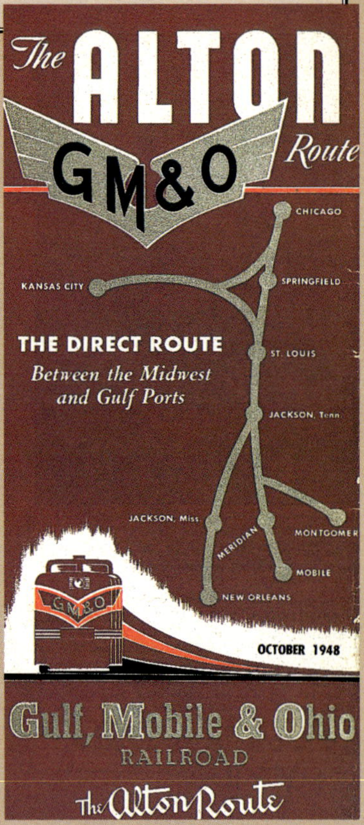


**O**ne of the nation's premier railfan trips in the 1950s was GM&O's Bloomington, Illinois to Kansas City gas electric and trailer. On a hot, humid June 4, 1957 George Billmyer and Jim Ozment (with the cigar) were photographed on the rear platform of the observation car or, more accurately, trailer 2551, at Roodhouse, Illinois.

North of East St. Louis was EMD territory on the GM&O. On January 3, 1956 General Motors E7s 101 and 101A lead the ABRAHAM LINCOLN north at Mitchell. The varnish is running on Big Four steel here as part of a shared trackage arrangement.

**P**owered by a matched set of Alco-GE PA1s 290 and 291, THE GULF COAST REBEL heads east from Union Station on May 21, 1958. After crossing the river and a brief stop at Relay Depot, THE REBEL will head south. Lines south of St. Louis were considered Alco territory on the GM&O.

**O**n a sunny March 28, 1959 GM&O's two Baldwin DR-6-4-20 passenger units, the 281 and 280 were stored at Jackson, Tennessee. These units were regulars on the REBELS until replaced by the Alco PA1s.

**A**lso stored at Jackson's Iselin Shops, the lightweight REBEL trains were awaiting their ultimate fate on that same March 28, 1959. Power cars 353, 354 and 352 sit with their respective trains out behind the shop facilities. Purchased by the GM&N (predecessor to the GM&O) in 1935, the same year Jim was born, these trains were used in St. Louis - New Orleans service for awhile.

As a child, Jim recalls taking the GM&O' subsidiary Gulf Transport bus to Relay Depot with his mother. There they would catch these strange looking little trains and ride to Jackson, Tennessee. At Jackson, they would transfer to the L&N for the last 27-miles to the family hometown of Lexington. On one trip the "Rebel" failed on the road and the train was rescued by one of the GM&O's worn out Pacifics.

These train sets were a joint effort by American Car & Foundry (ACF) and Alco-Westinghouse. Unlike contemporary articulated designs such as Illinois Central's GREEN DIAMOND or Burlington's early ZEPHYRS, these trains were made up of separate cars and power units.

# ILLINOIS CENTRAL RAILROAD

The Illinois Central has a large yard complex in southeast East St. Louis served by two mainlines. A line from Springfield, Illinois enters the area through Madison and Bridge Junction and through trackage rights over the TRRA, reaches "B" and "A" yards through Valley Junction. The southern route from Carbondale, Illinois enters the yard directly over IC trackage from the south east. Both lines combine to make what is essentially a St. Louis loop which connects to IC's Chicago Memphis mainline to the east. Several smaller yards, the "B" hump yard and the flat "A" yard were operated as the East St. Louis Terminal Division until 1958. After May 1958, the loop serving St. Louis and all terminal operations were combined into the St. Louis Division, stretching from Clinton, Illinois on the north to Carbondale on the south.

All terminal and road freight power were serviced at the "B" Yard facilities in East St. Louis. Passenger locomotives were usually serviced in East St. Louis, only rarely by the TRRA.

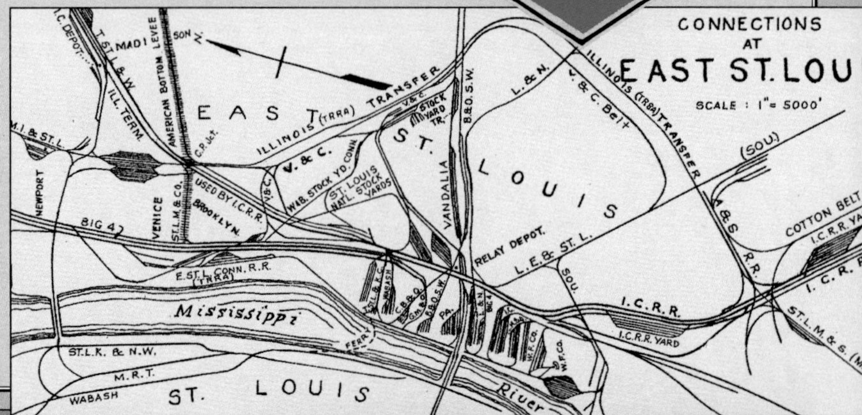

**A**n inbound freight off the Madison line arrives at East St. Louis on January 2, 1955. Jim photographed this scene from Illinois Terminal car 284 as it traversed the east approach trestle of the McKinley Bridge.

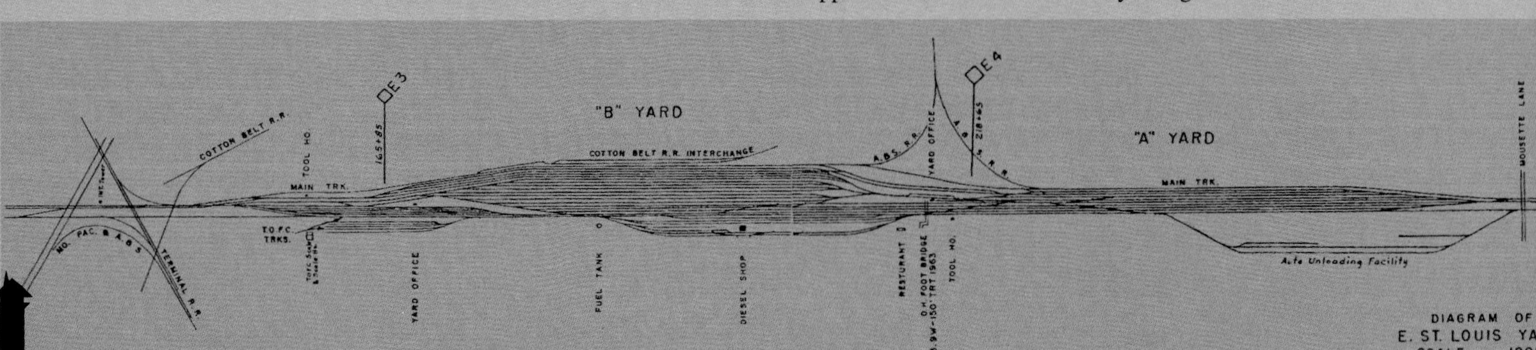

**M**ountain 2416 had just arrived from Carbondale, Illinois when photographed at East St. Louis on March 6, 1955. This class of power was used in both freight and passenger service until displaced by General Motors diesel locomotives.

**T**he Mikado or 2-8-2 wheel arrangement was one of the most popular freight locomotive on the IC. At its peak, this group had 517 engines in service. On March 6, 1955 Mike 2124 was inside for inspection at the East St. Louis roundhouse.

**A**fter arriving from Carbondale earlier that morning, Mountain 2614 was taken into the East St. Louis roundhouse. Although in good repair, steam locomotives like the 2614 simply required more attention and support facilities to do their job than an equivalent diesel unit.

**N**ovember 21, 1967 was late in the game for IC passenger service to St. Louis. One of five E8 passenger units rebuilt and renumbered by the Paducah shops prior to Amtrak, the 2022 leaves St. Louis headed for Carbondale. Though Tower One still stands, much of the trainshed and accessory structures at the mouth of Union Station have been removed by this time.

On December 30, 1954 Pacific 1155 was northbound when photographed at Relay Depot. Part of an order for 65 Pacific types purchased from Alco-Schenectady, this group handled mainline passenger assignments until replaced by the 4-8-2's. After downgrading to secondary passenger runs and freight service, the class lasted through the mid 1950s.

James L. Ozment is recorded at work on January 29, 1958. Beneath the focusing cloth of a 6 1/2 x 8 1/2 format view camera, Jim photographs Mike 1697 in the Carbondale dead line.

# Illinois Terminal Railroad

The Illinois Terminal Railroad began life as an electric interurban line between Danville and Westville, Illinois in 1901. Through the years it grew into a 400-mile system, the second largest in the United States.

It differed from most interurban lines in that it developed a substantial freight interchange with trunk line railroads. To handle this growing volume of freight traffic, the IT built belt lines around the major cities along its route through central Illinois. This served the dual purpose of providing better freight service to its on-line shippers and segregating high speed passenger traffic from low speed freight traffic.

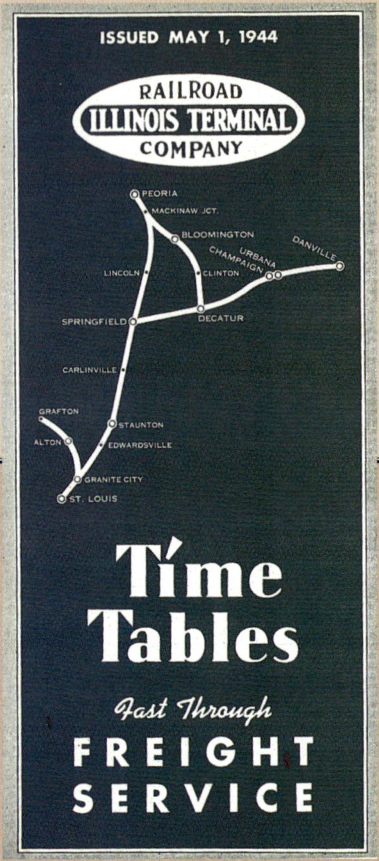


In its heyday the IT had three major routes: St. Louis to Peoria via Springfield (172 miles), Springfield to Danville via Decatur and Champaign (123 miles) and Decatur to Mackinaw through Clinton (66 miles). At one time the line offered Sleeping and Parlor Car service. After World War II passenger prospects seemed so good that IT purchased eight streamlined cars in 1948. They were the final interurban cars built for service in the United States, until the South Shore acquisition of new equipment in the 1970s.

In 1956, the Illinois Terminal was purchased by a group of eleven railroads. At the time of purchase, the IT still operated over 328 miles of main line and owned the McKinley Bridge, serving as its access to St. Louis proper. Through passenger service was cut back piecemeal, with the final scheduled Peoria-St. Louis run occurring on March 3, 1956. System wide passenger service ended in 1958, the same year the McKinley Bridge was sold to the city of Venice, Illinois. Miles of underutilized trackage were then abandoned in favor of using trackage rights over the lines of its owner roads.

Jim first encountered the IT on one of the family "Sunday Drives" in 1947, and was hooked. After he went to work for Carna Studio, planning and opportunity came together and he began riding and shooting the IT with a passion. He covered most everything between St. Louis and Springfield in one way or another after 1950. His Kodachrome sampling appears here.

Class B freight motor 1576 was at the Granite City Shops on September 19, 1954. Eighteen of these units were built by the IT at the Decatur Shops between 1910 and 1918.

Car 471, a former East St. Louis and Suburban car, was at Granite City on September 19, 1954. This car was used on the local run between Granite City and St. Louis.

Line car 1702 and snow sweeper 019 were also parked at Granite City on September 19, 1954. On weekends, photographers could be certain of finding most of the roster tied up at the Granite City Shops, and at least some equipment parked where it could be easily photographed.

**G**ranite City local cars 454 and 471 wait for the Monday morning call at Granite City. These shops were located at 15th and Madison Streets in Granite City, Illinois.

**I**llinois Terminal passenger service was not limited to only those vehicles with flanged wheels. A group of GMC buses, with Unit 93 in the foreground, are idle on Sunday, September 19, 1954. The buses provided connecting service to East St. Louis.

**C**ombination car 273 was built by St. Louis Car in 1913. Wearing the standard Traction Orange and Tuscan Red paint scheme introduced in 1924, the forty-two year old veteran is at Granite City on August 23, 1955. Immediately behind the 273 is a car repainted in the 1948-inspired blue and silver scheme.

**O**n August 23, 1955, Class C motor 1587 was parked at Granite City. Compare the size of the 1587 to the green Class B motor parked behind it.

PCC car 452 rolls through down town Madison, Illinois on September 11, 1955.

Alco-GE S-2 703 works industries along Eleventh Street in St. Louis on September 11, 1955. This location is near the entrance to the Illinois Terminal depot.

Northbound car 301 is on the approach to the McCambridge Avenue Bridge on February 22, 1956. The location is near Madison, Illinois. Combination cars 300-302 were built by St. Louis Car in 1948.

The system map of the IT was displayed above the street entrance to the St. Louis Depot at 710 North 12th Street. It is now on display at the National Museum of Transport at Barretts Station.

Southbound car 302 crosses the McCambridge Avenue Bridge on February 22, 1956.

Stainless steel car 352 brings up the rear in this consist at Twelfth Street Terminal in St. Louis. The terminal was a subway beneath Washington Street. Car 352 was built as a reserved seat coach with a buffet section in 1948. It was named *Pierre Laclede*, for the French entrepreneur who settled the area prior to Lewis and Clark. The date was March 4, 1956, the day after the final Peoria (Illinois)-St. Louis through passenger run.

PCC car 453 approaches the end of her run at Twelfth Street, St. Louis on March 4, 1956.

Cars 470 and 473 are running in multiple as they roll along 17th Street in Granite City on March 9, 1956.

**P**CC car 450 waits along with much newer and larger car 300 in the Granite City Yards on March 9, 1956.

**C**lass C motor 1595 sits at the Granite City shop on March 27, 1956. Beneath the grime she wears the solid green freight scheme. Built in the company shops during the 1920s, General Steel Castings of Granite City provided cast frames for the Class C motors. Trucks were salvaged from retired passenger equipment. Car bodies for the Class C motors were nearly identical to those used on the Class B motors and fabricated in the company shops. Ready for service, these units weighed 80-tons each.

**T**his line up at Granite City on May 1, 1956, included PCC, Granite City Local and Alton Local cars.

Do not look for this industrial road in your *Official Guide* because you will not find it there. This little 0-4-0 worked at a quarry 2-miles west of Pacific, Missouri near the Frisco right of way. These views of No. 4 going about its daily routine illustrate a phenomenon observed at the end of the steam era. Industrial sites with second-hand locomotives were sought out and venerated by fans due to the working steam locomotives on the property. Such was the case at this location west of St. Louis, photographed on December 30, 1957.

# LOUISVILLE & NASHVILLE

Cutting across southern Illinois from Evansville, Indiana, the L&N enters East St. Louis from the east, south of the B&O. Third Street Yard was L&N's primary yard and engine terminal, located east of Relay Junction and the National Stock Yards. Lower Yard, on the river front, connected to the TRRA on the Wiggins Ferry line and was just south of the Eads Bridge.

While L&N freight power never ventured west of Eads Bridge, its passenger power was serviced at TRRA's Fourteenth Street facility. According to TRRA employees, the L&N was the last road being serviced at Fourteenth Street. When an L&N unit was delivered, craft people from TRRA's Brooklyn Shops would drive over and take care of it. This was after the other eastern roads had transferred maintenance to their own facilities on the gateway side of the river.

East St. Louis was the end of the line for this batch of Mikados. A final run to the steel mills at Granite City behind a transfer locomotive was all they were capable of after the main rods, lights and other appliances had been salvaged. On September 19, 1954 Mr. Ozment mugs for the camera from the cab of the 1527 at East St. Louis.

# Missouri-Kansas-Texas

The MKT, or "Katy" as it was widely known, ran into St. Louis over a single track line. Coming due east from Parsons, Katy's line joined the CB&Q line from Hannibal at Machens, Missouri. The joint track ran east along the Mississippi to West Alton and the junction with the Missouri & Illinois Bridge & Belt Railway. Turning south at West Alton, the line crosses the Missouri River at Fort Bellefontaine and continues south and east to the west bank of the Mississippi and Baden. At Baden, the Katy leaves the Burlington's line for its own terminal, Baden Yard.

Interchange of freight was done via transfers from TRRA into Baden Yard. By the 1950s passenger operations into St. Louis were limited to the MKT/SLSF (Frisco) joint operation of THE TEXAS SPECIAL. While Katy passenger units were used on this train, they were serviced by the TRRA at Fourteenth Street. This train ran west over the Frisco to Kansas City. South of there, THE TEXAS SPECIAL was a Katy operation to Dallas and San Antonio, Texas.

Assigned to Baden for switching and local work, the 1576 idles on January 2, 1956. Baldwin AS16's by birth, the 1576 and sisters would be re-engined by EMD by 1960, making them GP9m's.

Adjacent the engine terminal at Baden was a modest trailer loading facility. On January 2, 1956 a trio of cars awaits movement west. Circus style, roll-on/roll-off, loading was standard operating procedure in the 1950s when this service was just beginning.

An office was needed for the crews and clerks who loaded the piggyback cars, so the Katy pulled this veteran outside-braced caboose from revenue service. Serving as the office for the piggyback facility, the 370 was recorded on January 2, 1956.

# Missouri Pacific Railroad

The Missouri Pacific had, by far, the most extensive operations of any Class One road in the St. Louis/East St. Louis area. Three mainlines and numerous branches connected the outside world to St. Louis via the Mopac. Early exposure to the Mopac from his home in South St. Louis allowed Jim to learn its operations and equipment and made shooting different aspects of the Mopac an easy pastime.

On the east side of the Mississippi River, the Chester Subdivision of the Illinois Division ran south from Valley Junction to the yard complex at Dupo, a large flat yard and engine terminal. In addition to collecting transfers from East St. Louis, Dupo also received coal traffic from subsidiary Missouri-Illinois operations in the southern Illinois coal fields. Trains were dispatched south to the Mississippi River crossing at Thebes, Illinois and then on to Poplar Bluff, Missouri. St. Louis-Southwestern (Cotton Belt) traffic had trackage rights over this line south of Valley Junction to Illmo, Missouri, where their trackage starts. At one time, a car ferry to West Ivory, southwest of Dupo was served via a short branch line from the yard, but that has long been superseded by trackage rights over the MacArthur Bridge through Valley Junction.

The St. Louis Terminal Division encompasses trackage north from Dupo through Valley Junction and over the MacArthur Bridge to the Mill Creek Valley. This includes industrial trackage south along the river front. South of the MacArthur Bridge, Terminal Division included the trackage from Lesperance Street to Davis Junction, the Lesperance Sub. West through the Mill Creek Valley, Terminal Division trackage includes Mopac's 23rd. Street Yard and the Ewing Avenue engine terminal west of Union Station and the various junctions and interchanges at Grand Avenue and Iron Mountain Junction.

West of Maplewood, the Sedalia Sub of the Northern Division is the mainline west to Jefferson City and Kansas City. Just east of Kirkwood, the Carondelet Sub heads south and east. This line once served as an outer belt line for the MP, bypassing the congestion of the Mill Creek Valley. At Iron Mountain Junction, the DeSoto Sub of the Arkansas Division branches to the south. After cutting across south St. Louis, this line joins the Carondelet line of the St. Louis Terminal Division and parallels it a few blocks southeast to Davis Junction. There the DeSoto Sub turns south to follow the river a short distance before climbing to Iron Mountain and Tip Top on the way to Poplar Bluff.

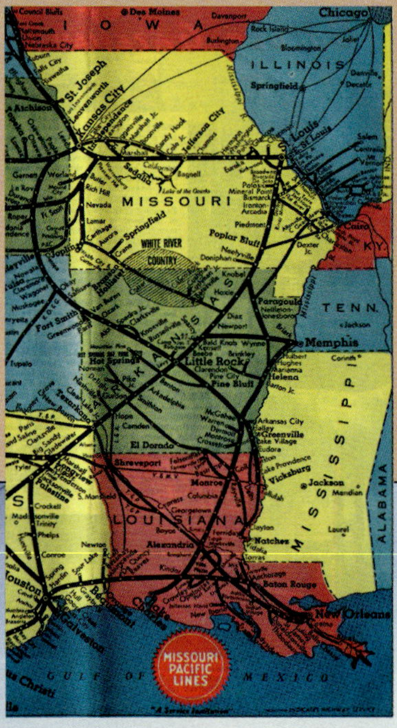


In addition to its physical presence in the St. Louis area, Mopac insured that the public was aware of opportunities to use its services. Looking north along the 3900 block of Kingshighway, there are several reminders to consider here. Ignoring the wayward Soo Line boxcar in the middle of the road (an escapee from a local switch crew) the overpass promoting the Eagle Fleet frames the scene and grabs your attention. In addition, there is the billboard to the left to make sure the subconscious got the message. Jim recorded this scene on October 15, 1954.

In addition to receiving and dispatching St. Louis traffic for the Illinois Division and servicing motive power, Dupo served as the bone yard for retired equipment waiting for disposition. This was true during both steam and diesel eras. On December 30, 1954 still handsome Mikado 1417 silently waits for scrapping.

THE TEXAS EAGLE eases to a stop just west of the Fourteenth Street Viaduct on September 2, 1954. All trains using St. Louis Union Station backed into their respective tracks on the multi-tracked wyes that made up the approaches to the depot. In this case, all Missouri Pacific trains departed as westbounds, so PA1 8002 has pulled east of the station in order to back in and face west for departure. The stainless steel Eagle emblem really stands out on the dark blue nose.

**O**n May 19, 1956 THE TEXAS EAGLE, train No. 22, passes Tower One at the throat of Union Station. Today's motive power is a mix of EMD E8 7021 and Alco-GE PA2 8017. Once No. 22 departs, the train will proceed west to Iron Mountain Junction, just west of Tower Grove. There it will head south on the Oak Hill Branch of the DeSoto Subdivision for the trip to San Antonio.

**P**assing TRRA's Fourteenth Street engine facility, PA1 8002 briefly pauses next to two Pennsy E units waiting to back down to Union Station. Two tracks to the south, some Baltimore & Ohio units heads east to their own servicing facility at Cone Yard on the other side of the Mississippi. This busy scene was recorded on September 2, 1954.

Ewing Avenue Engine Terminal serviced the passenger power for all Mopac trains using Union Station. Power for freight trains dispatched out of 23rd Street Yard was also serviced here. Alco-GE 380 is first out on the set of FAs at the sanding tower. The terminal was photographed by Jim from the Jefferson Avenue Viaduct on September 24, 1956.

# *Condensed* FREIGHT SCHEDULES

***Between the St. Louis Gateway and . . .***

| *Read Down* | DALLAS, FT. WORTH, EL PASO and LOS ANGELES | | *Read Up* |
|---|---|---|---|
| Train 65 and Connections | Train T&P 67 and Connections | Train 60 and Connections | Train 72 and Connections |

Headed east from Ewing Avenue, a matched set of Alco-GE FAs roll beneath Jefferson Avenue to go pick up their train. The trailing unit, which will actually lead the train west, is FPA2 391. These units were equipped with extra water capacity and steam generators for use in passenger service as the need arose.

Looking east from the Grand Avenue Viaduct, on May 31, 1955 a westbound freight is getting underway with FPA2 364 on the point. The tracks to the left of the locomotive are Wabash, Frisco and TRRA Terminal District.

| | | |
|---|---|---|
| M. T.)....Lv | .......... | 8.35PM Su |
| T.).......Lv | .......... | 2.00AM Mo |
| ..........Lv | .......... | ®© |

**H, LINCOLN and OMAHA**

| | Train 70 and Connections | Train 70 and Connections | Train 62 and Connections |
|---|---|---|---|
| ..........Ar | 6.30AM Tu | 6.30AM Tu | 6.00PM Tu |
| ..........Lv | 8.30PM Mo | 8.30PM Mo | 10.00AM Tu |
| ..........Lv | 1.30PM Mo | 5.00PM Mo | 11.50PM Mo |
| ..........Lv | .......... | 4.00PM Mo | .......... |
| ..........Lv | 10.00PM Su | .......... | .......... |
| ..........Lv | 6.00AM Mo | .......... | 6.30PM Mo |

**TOPEKA and SALINA**

| | Train 62 and Connections | Train 80 and Conn. |
|---|---|---|
| ..........Ar | 6.00PM We | 3.00AM We |
| ..........Lv | 1.30PM Tu | .......... |
| ..........Lv | .......... | 2.15PM Mo |
| ..........Lv | .......... | 10.00AM Mo |
| ..........Lv | .......... | 3.00PM Mo |
| ..........Lv | 6.30AM Tu | .......... |

**E, NEW ORLEANS and LAKE CHARLES**

| | Train 370 and Connections | Train 360 and Connections |
|---|---|---|
| ..........Ar | .......... | .......... |
| ..........Ar | 3.00AM We | 11.30AM We |
| ..........Lv | 9.00PM Mo | 8.30PM Tu |
| ..........Lv | 2.55AM Tu | .......... |
| ..........Lv | 5.00PM Mo | 9.20AM Tu |
| ..........Lv | 2.00AM Mo | 9.00PM Mo |
| ..........Lv | 6.30AM Mo | .......... |
| ..........Lv | .......... | 9.30AM Mo |

*not guaranteed*

Grand Avenue Tower controlled the interlocking plant protecting interchange and mainline movements for the Wabash, Frisco, TRRA and Mopac. On May 31, 1955, a rare EMC SW, the 9003, works eastward at Grand Avenue with a Swift wooden reefer in tow.

Shooting from the Ewing Avenue viaduct, Jim got a good down-on shot of Texas & Pacific E7 2000 on September 24, 1956. Among the details visible from this angle are the T&P herald on the painted Eagle emblem on the nose and the white roof, compared to the solid blue roof on Alco passenger units.

**T**ower Grove Station was a stop for both eastbound and westbound passenger trains. Just west of the depot was Iron Mountain Junction and a diamond where the Mopac Oak Hill Branch crossed the Frisco at grade. On May 9, 1958 the eastbound TEXAS EAGLE, lead by PA2 8010, has just come off the Oak Hill line as it slows to let passengers for west St. Louis detrain.

**C**ompare two different views of the Tower Grove Station recorded from the Vandeventer Viaduct. The slogans and herald atop the station were more impressive after sundown. The timed exposure was created on September 17, 1954. Looking northeast on May 9, 1958 the Missouri Pacific mainline disappears around a curve, approaching Grand Avenue and the Mill Creek Valley corridor.

All Trains
TO AND FROM ST. LOUIS
Stop at
TOWER GROVE STATION
● A centrally situated Missouri Pacific Station, maintained for your convenience — to save time at the start or the finish of your journey.
● Tickets sold, Pullman reservations held and baggage checked for Tower Grove passengers.
● Street car and bus line service from all directions — a regular stop for Sarah Car (No. 42) and Lindenwood Bus (No. 92).
● Taxi fares less from almost every section of St. Louis.
● FOUR OTHER CONVENIENT NEIGHBORHOOD STATIONS:
Broadway, Maplewood, Webster Groves, Kirkwood
MISSOURI PACIFIC LINES
WEST - SOUTHWEST
ON VANDEVENTER
HALF-BLOCK EAST

MISSOURI PACIFIC LINES
ROUTE OF THE Eagles
WEST-SOUTHWEST

**Y**ears earlier MP operated suburban service for the western suburbs of St. Louis. As the Sedalia Subdivision climbed out of the Mill Creek Valley, each location received a depot befitting its size and importance as a source of revenue for the Missouri Pacific. One commuter stop, within the St. Louis City limits, was Ellendale. At the intersection of Southwestern Avenue and West Main, this stop had only a shelter for waiting passengers. Though no longer used, the shelter was in good repair on March 1, 1955 when Jim shot this scene.

**W**estbound at Ellendale, PA2 8018 and mates roll through a light dusting of snow into the diffused winter light of February 13, 1958. Track maintenance workers momentarily stop to give the train a roll-by inspection. This is more than a friendly gesture on the part of the track workers. They examine a moving train from different view points and can relay any potential problems by hand signals to the crew.

When wages were low and help plentiful, grade crossing protection in urban areas was provided by crossing watchmen. Depending on the location, some shanties were simple sheds at trackside while others were more elaborate. This elevated shanty at Machlind and Manchester Streets in west St. Louis is one of the latter. Photographed in February 1955, this structure granted the watchman increased visibility since the tracks approached on a sweeping curve from the west. Operators would be alerted via phone signal of an approaching train and would activate the hydraulic lever that lowered the gates. Once the train had passed, a relief valve was turned and the gates raised. These shanties and the operators were eliminated when the crossings were automated in the late 1950s.

At that same crossing, though seen here from the east, the COLORADO EAGLE rounds the curve, protected by the crossing watchman. The motive power displays some details unique to the Missouri Pacific. Both the 7010A and second unit are EMD E7s, identified by the wire covered air intakes along the upper sides of the car body, below the roof. Instead of the usual rectangular windows found on this model, the Mopac has opted for portholes along the sides.

A local freight works the eastbound main in west St. Louis on March 15, 1956. Wearing the original black with white lettering switcher scheme, Alco-GE S2 9113 was photographed near Knox Avenue.

**A** plume of steam rises in the cold morning air of February 13, 1958 as pile driver X1032 works near Southwest Avenue. The location is near the St. Louis city limits adjacent the suburb of Maplewood.

**K**irkwood, Missouri is at milepost 13.4 west of St. Louis. Kirkwood is the summit of the grade for eastbound trains. During the steam era, this was actually where helpers cut off after pushing from the west. About one mile east of the depot is Kirk Junction. There the Carondelet Sub leaves the Sedalia Subdivision and runs south and east to parallel the Oak Hill branch to Broadway Junction. On August 30, 1954 an eastbound freight has just crested the summit and, passing the stone depot, begins its descent to the Mill Creek Valley.

Barretts, Missouri is west of Kirkwood and east of Keyes Summit. It is on the eastward climb to the summit at Kirkwood. Terrain here markedly changes from the flat bottom lands characteristic of the Mississippi Valley to the more rolling country typical of the Ozarks to the south and west. Originally the mainline at Barretts passed through two single track tunnels about one-half mile apart. These were bypassed with cuts during Would War II which allowed double tracking and greater clearances and of course, eliminated the problems of smoke within confined bores for crews and passengers. Barretts is the name used in Missouri Pacific timetables. Since many people know the location as Barretts Station, that is the name we use here.

On March 30, 1954 two freights meet on the curve west of Barretts Station. Missouri Pacific was quite the Alco stronghold in the early 1950s as the two solid FA consists illustrate. The westbound is lead by FPA2 391, one of the FA types equipped with a steam generator.

Another westbound, this one lead by FPA2 389, at West Barretts Station on November 4, 1954.

Here is the cut which replaced the west tunnel. Lighting here favored photographing westbound trains and this view is no exception. A standard FA2, the 379, and two mates roll west through Barretts Station on September 13, 1954.

Compared to the spacious clearances provided by the new cut, the west tunnel at Barretts Station seems absolutely claustrophobic. On September 13, 1954 Jim photographed the west portal of the west tunnel and cut. At the far end of the cut are the grounds of the National Museum of Transport, established on the west side of the Mopac between the two hills at Barretts Station in late 1946.

Barretts Station was located at mile post 16.25 west of St. Louis.
The depot was relocated to this site after the line change and bypassing of the tunnels.

Running left-handed, on August 30, 1954 local train No. 35 departs Barretts Station westbound. This train was a commuter local from St. Louis to Pacific, Missouri.

Working the westbound local freight, No. 91, Alco-GE RS3 4506 switches Barretts Station on June 2, 1955. In the late 1950s, Missouri Pacific made drastic changes within its Motive Power department after the promotion of John T. German to the Vice-President Motive Power position. Being completely dedicated to EMD's 567 diesel engine, J.T. German directed that all non-General Motors power either be scrapped or re-engined with that standard power plant. Consequently, the RS3 fleet was equipped with 567 diesel engines and the unique "GP12" fleet created. More unusual was the re-engine work done on almost new Alco RS-11 road switchers. After only a year or so of service, they received 567C power plants to create the "GP16" road switcher.

With the early Museum of Transport collection serving as a backdrop, GP7s 4185 and 4140 lead local 91 west on January 20, 1955.

From a raised vantage point, the same basic scene has No. 91 heading west behind GP7 4169 on August 9, 1954.

Keyes Summit, milepost 17.1 west of St. Louis, is where the main lines of the Missouri Pacific and the Frisco briefly converge as they leave the Mississippi Valley. On March 10, 1957 eastbound Extra 358 passes beneath the signal bridge working upgrade towards Kirkwood. The Frisco mainline is a few hundred feet to the left looking to the west.

Just west of the Tower Grove Station is Iron Mountain Junction, named for a Missouri Pacific predecessor, the St. Louis, Iron Mountain & Southern. The Oak Hill branch of the DeSoto Sub heads south through suburban St. Louis to Broadway Station. This line was primarily a passenger line, with a local way freight being the only scheduled freight traffic. Equipped with Automatic Block Signals (ABS) the line was operated as a double tracked mainline. On May 22, 1958 PA1 8006 leads two other units and train No. 25 near Loughborough Avenue southbound for Poplar Bluff.

The double-tracked Oak Hill Branch passes through Carondelet Park on the south side of St. Louis before converging with the Carondelet line at Broadway Junction. Looking north through the Park on the foggy morning of January 5, 1958 this view shows a bit of the neighborhood where Jim Ozment grew up watching trains.

The crossing shanty at Bates Street was undergoing automation by the signal department when Jim recorded this view in late March 1955. When in grade school, Jim befriended the second shift crossing operator at Bates Street, Stella Rueprich, and after watching and assisting in the crossing gate operations countless times, came to be trusted to perform the operation himself and then ring the next operator down the line signaling the approaching train. Not a bad assignment for a kid on his way home from school, particularly when the afternoon lineup had northbound No. 4 at about 4:15, and southbound No.21 (MP TEXAS EAGLE), No. 1 (T&P TEXAS EAGLE) and No. 31 (SUNSHINE SPECIAL) between 5:30 and 6:00 p.m.

At Broadway Junction, the Corondelet Sub turns north and, via the Stein Street connection, joins the Lesperance Sub, running north along the river towards downtown St. Louis and Lesperance Yard. Turning south, the DeSoto Sub passes through Davis Junction, and its connection with the Lesperance Sub, to the yard and former car ferry at Ivory. Heading southwest, the line pulls away from the Mississippi as it climbs out of the river valley. Crossing the Meramec River further south, Train No. 25 heads for Poplar Bluff on November 25, 1954.

The summit of the climb south from St. Louis is near this cut at Tip Top, Missouri. Here, northbound train No. 4 crests the grade on June 15, 1958. Train No. 4, lead by the 7019, left Little Rock at 8:25 a.m. and will be in St. Louis at 5:15 p.m. Connections will be made there to go east on the ST. LOUISAN, Pennsy train No. 32 at 6:25 p.m.

# Nickel Plate Road

The New York, Chicago & St. Louis Railroad, more commonly known as the Nickel Plate Road, traversed south central Illinois from the Junction at Frankfort, Indiana. This portion of the Nickel Plate was formerly known as the Toledo, St. Louis and Western, or Cloverleaf. Entering the area from the northeast, Nickel Plate ran parallel to the L&M (C&NW) and Illinois Central from Edwardsville and Glen Carbon to its yard in Madison. Freight was interchanged via the TRRA Merchants Bridge line and the A&S outer belt to Madison. A small yard along the riverfront received East St. Louis interchange from Wiggins Ferry Yard, which was then transferred to Madison via Bridge Junction. Passenger trains, the unnamed Numbers 9 & 10, ran via Bridge Junction and the TRRA Venice Line to Relay Junction and the Eads Bridge.

Northbound at Granite City, a fast freight trailing a wood caboose leaves the Mississippi Valley in its wake on March 27, 1956. Wearing the "Nickel Plate High Speed Service" slogan and distinctive yellow marker flags, the caboose passes to the west of Horse Shoe Lake.

Recorded from the Eighteenth Street viaduct, PA1 180 heads train No. 10 out the east leg of the wye at Union Station on May 21, 1958. Running east to Twelfth Street, No. 10 will head north through Eads Tunnel and Bridge to Relay Junction, then running north to Bridge Junction and home rails. Though the road focused on freight service, the immaculate appearance of unit 180 reflects the comments of TRRA employees who claimed that Nickel Plate power was among the cleanest they serviced at Fourteenth Street.

REFRIGERATOR
RAILWAY EXPRESS
AGENCY
RFX 7404

# PENNSYLVANIA RAILROAD

The Pennsylvania Railroad enters East St. Louis from the east, the line originating at Indianapolis, Indiana and running southwest through Terre Haute. From Terre Haute, the Pennsy runs straight to East St. Louis, on a more direct and southerly route than rival New York Central's Big Four line. At Caseyville, just east of East St. Louis, the tracks of the Pennsy and the B&O come together and parallel each other all the way to Relay Junction.

Rose Lake Yard is the main yard and engine terminal located between the A&S outer belt line and the TRRA Merchants belt line. The yard is located on the edge of Fairmont City, Illinois. Passenger locomotives and rolling stock were serviced by the TRRA in St. Louis. Freight was moved via transfers on the two terminal roads.

A classic set of pin-striped E7s, lead by the 5869, round the wye at Eighteenth Street on March 21, 1958. These units were serviced by the TRRA at Fourteenth Street and will power the eastbound ST. LOUISAN, train No. 32.

In September 1954 the 5791 and 5707 wait at Fourteenth street for the call to Union Station. These E units will power train No. 30, THE SPIRIT OF ST. LOUIS. The TRRA service facilities are immediately to the right of the units.

A variety of Pennsy passenger units await servicing at Fourteenth Street on December 8, 1957. This view is to the west from the Twelfth Street Viaduct. Surging Holiday traffic is the cause for the abundance of motive power. Brunswick Green 9857 and mate are dual service FP7s waiting for a call on an eastbound mail extra.

Very late in the Pennsy era, spartan E8 4306 and a single-striped booster back down to Union Station. It is November 1967, and the consolidation into the ill-fated Penn-Central is only a few months away.

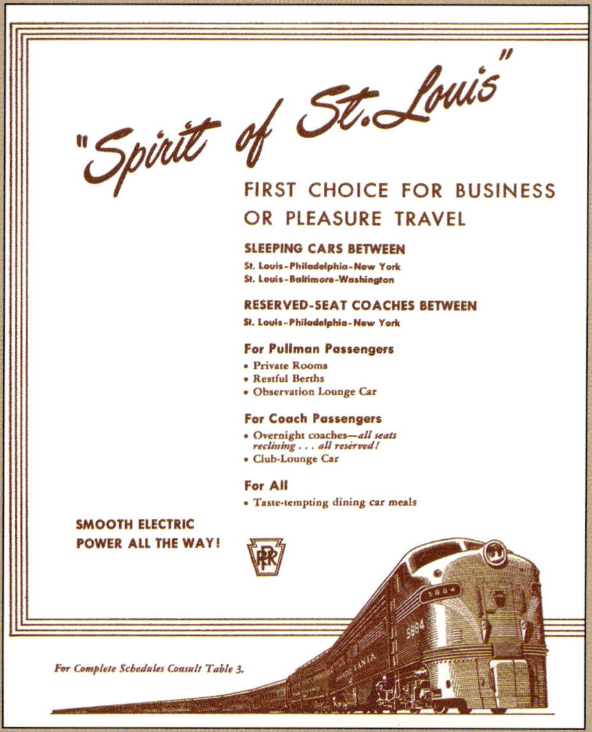

**A** TRRA transfer heads east as Pennsylvania train No, 13, THE MAIL AND EXPRESS, heads west after stopping at Relay Junction on June 4, 1956.

# St. Louis Public Service Company

Saint Louis Public Service Company penetrated all parts of the city and its suburbs in the 1930s and a young Jim Ozment spent his childhood riding the Public Service cars. His mentor was his grandmother, Mary Bette Miller, described as a "Boomer" who never learned to drive. "Mom" Miller treated the Public Service System as her personal transportation and knew how to get everywhere on it. On Sundays she would take Jim in tow and go streetcar riding. Sometimes the goal was to visit friends, sometimes just find new lines to ride.

When a new PCC car was displayed to the public in 1940, "Mom" Miller took 5-year old Jim down to see it. By the time Jim started taking slides, only a few lines were left, all operating PCC cars. Nevertheless he recorded all he could. Having cut his photographic teeth in such an environment, it is no surprise that Jim Ozment was a traction fan then and remains a passionate supporter of urban light-rail operations today.

Public Service lines traversed most of the main arterial routes within the city. As was the case with other street car operations around the country, this service came to its premature demise at the hands of the automobile and petroleum companies and a short sighted public. An appropriate irony is that St. Louis is in the process of building a light-rail system today, serving much of the area once served by the Public Service Company and its Illinois-bank counterparts.

**A** line up of PCC cars at the PSC shops at Spring and Park on March 3, 1955. Left to right are cars 1597, 1585, 1514 and 1767. These cars were built by St. Louis Car in 1940, except for the 1767, which came in 1945.

**A**t the Park Avenue Cinder Yards, freight motors 97 and 99 are parked for the weekend on March 3, 1955.

Street trackage requires constant maintenance and PSC had a specialized fleet for that purpose. At the Spring and Park material yard in March 1955 rail grinder 215 and weed sprayer 60 wait for the call for service.

Out on line, overhead line maintenance car 160 was at 39th and Park Avenue on January 13, 1955.

The Wellston Loop depot contained a small diner in its lower level. This was the appearance of the depot on March 31, 1959.

Saint Louis is far enough south so while it does occasionally snow, the snow usually does not stay on the ground very long. Taken during a late spring storm, here are PSC cars lined up at the intersection of Chippewa and Grand Avenue on March 25, 1955.

The PSC system was being reduced in size and scope by the mid-1950s and surplus equipment sold off to other buyers. On March 3, 1955 PCC cars 1555 and 1564 were on flatcars at the PSC / Frisco interchange waiting for movement east. The two fifteen year old cars were on their way to a new owner, the city of Philadelphia, Pennsylvania.

On September 26, 1957, PCC car 1744 was on the Suburban Gardens loop on the Hodiamont line, in west central St. Louis. This line is near Washington University, Jim's alma mater.

# St. Louis Water Works Railway

This little line serves the city Water Works on the west bank of the Mississippi near the Chain of Rocks Dam northeast of downtown. The trolley operation came to an end in April 1955, but the line continues to operate using a small diesel electric to deliver cars of chemicals and other supplies to the Water Works.

Baden Station, near MKT's Baden Yard, was the terminal for WW Ry. Interchange of cars was done by the TRRA off the Merchants Bridge line. Running parallel to the MKT and CB&Q, the WW Ry. follows the river northwest from Baden Station. It crossed over the CB&Q on a trestle/plate girder bridge to avoid crossing at grade.

**C**oach 10 at Baden on April 5, 1955. Two cars were needed to transport employees at the necessary times of day, each starting at opposite ends of the line.

**C**ars 13, 11 and 10 parked at Baden on April 29, 1955. Employees would ride St. Louis PSC to within a few blocks of Baden and then walk to this station. The passenger coach would run out to the Water Works at the beginning or end of the assigned work shifts, transporting employees.

**T**he survivor, Diesel No. 3 pulls several cars of supplies along Riverside Boulevard on April 16, 1958.

# St. Louis - San Francisco

FRISCO

The SL-SF or Frisco enters the St. Louis area over two different routes. One, the Memphis Division enters St. Louis from the south. The other line, the first Sub of the Springfield Division enters from the southwest, i.e. Springfield, Missouri. Both lines are basically single track, except within the St. Louis Terminal area.

Approaching St. Louis along the west bank of the Mississippi, the Memphis line runs north through the suburbs of Mattese and Lemay. After crossing the Mopac's Carondelet Sub, the line runs along the western edge of the city to Southeastern Junction at the west end of Lindenwood Yard.

The Springfield Division approaches from the southwest passing through Keyes Summit and Kirkwood, where it also crosses Mopac's Carondelet Sub. It then passes through the suburbs of Oakland, Webster Groves and Shrewsbury. At Southeastern Junction, the line joins the Memphis Division before entering Lindenwood Yard, the major terminal in the area. Continuing northeast from Lindenwood, the line crosses the MP DeSoto Sub just west of Tower Grove Station. East of Tower Grove, Frisco and Mopac parallel each other to Grand Avenue Tower. After crossing the MP at Grand Avenue, the line continues to Ewing Avenue Yard where it ends. East of Twenty-Third Street, the east leads to Ewing Avenue Yard. Passenger operations continue east over TRRA Terminal District trackage to Union Station.

Close behind the Mopac in appeal, Jim would enjoy operations at Lindenwood from the Arsenal Street and Fyler Avenue viaducts. First, under the watchful eyes of his parents and later on his own, Jim would record the operations from Fyler Avenue. In Jim's opinion, the removal of the old Fyler bridge deprived photographers of one of the best camera platforms in the St. Louis area.

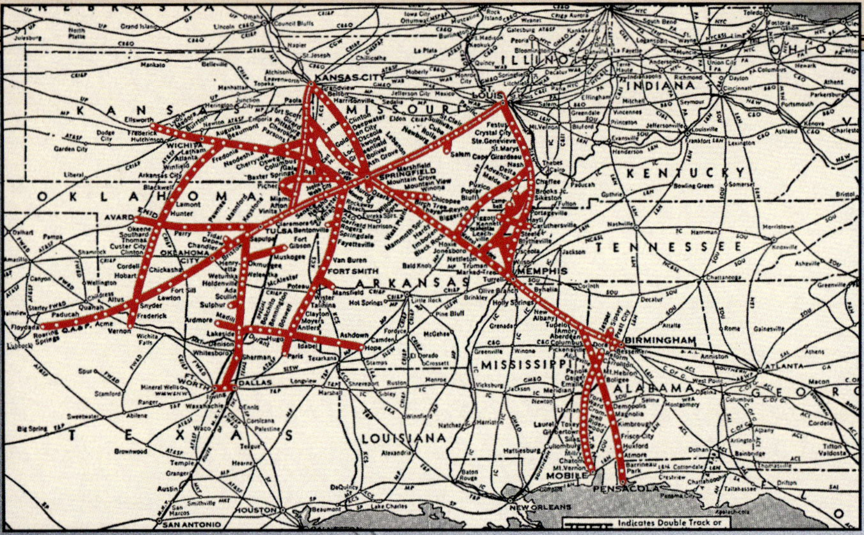

Near Grand Avenue Tower, NW2 253 and transfer caboose 70 are on a caboose hop between Lindenwood and Ewing Avenue. The Frisco line crosses the Wabash and Missouri Pacific at this interlocking plant. This move was recorded on September 4, 1960.

In the fading light of April 26, 1956 Train No. 1, THE TEXAS SPECIAL, starts to accelerate away from its stop at Tower Grove Station. A joint Frisco/Katy operation, today's train is lead by MKT E8 132. Crossing the Frisco mainlines in front of Katy 132 is the Missouri Pacific Oak Hill Branch. Iron Mountain Junction, where the Oak Hill Line diverges from the Sedalia Sub mainline is left of the Frisco tracks in this scene.

**F**risco's Tower Grove Station, as seen in a time exposure on September 17, 1954. This was shot from the Vandeventer Avenue Viaduct.

**A**n eastbound mail train slows for a stop at Tower Grove on May 9, 1958. This train is led by SL-SF E8 2003, a rebuilt E7 named *Steel Dust*. All the Frisco's EMD E-units were named for racehorses that won the Kentucky Derby. The MP mainline is seen above the train and Iron Mountain Junction is to the right.

**F**risco's Tower Grove Station was a multi-leveled affair, as seen here. Pedestrian traffic could enter from Vandeventer Avenue or the parking level on the north side. Once inside, all had to descend one more level to the platforms. On May 9, 1958, the eastbound TEXAS SPECIAL is about to leave for the final leg of the trip to St. Louis Union Station.

**E**astbound No. 2 accelerates around the curve east of Tower Grove Station on May 9, 1958.

**T**wo sisters await their end on August 22, 1954. Behind the Lindenwood roundhouse, the 4518 and 4502, both 4-8-4s, were parked on the garden tracks around the turntable.

**M**oved from the roundhouse, the 4517 was parked with the tender from the 4501 on August 29, 1954.

This surreal scene was recorded on August 29, 1954. Cabs of retired steam locomotives curve into the distance around the turntable whisker tracks at Lindenwood. Separated from their tenders, all the cabs have been carefully boarded up pending an unlikely call to duty.

Parked in the Lindenwood roundhouse on June 28, 1955 Mountain 1501 was still in excellent condition. Even as stand-by power, the Frisco kept these locomotives in good repair.

Lindenwood engine terminal became the final assignment for much of the steam fleet. On August 22, 1954 SL-SF 4517, a 4-8-4, was stored in the Lindenwood roundhouse prior to its retirement.

The standard face of railroading in the 1950s was that of the EMD F-unit. Parked alongside the diesel house at Lindenwood, F7 5035 typified the units that flooded rosters from one end of the country to the other and washed away all that was steam.

Threatening gray-blue skies of December 1955 hang over this westbound, preparing to get under way at Lindenwood. In the weak light of a winter day, three GP7s led by the 556 are in charge. The largest class of diesels rostered by the Frisco, GP7s totaled 128 units.

**L**indenwood by night was photographed on August 30, 1960 from Fyler Avenue.

**S**hot from the Fyler Avenue overpass near Lindenwood Shops, E8 2021, *Gallahadion*, rolls train No. 808 north on March 7, 1955. After a 7:40 a.m. departure from Memphis, the heavyweight SUNNYLAND keeps a leisurely pace northward to arrive at St. Louis at 3:20 p.m.

**A**n inbound freight rolls to a stop in Lindenwood on August 22, 1954. Lead unit 5051 is one of six FP7 units. Purchased for passenger protection, the FP7 is four feet longer than a standard F7, and is equipped with an increased water capacity and a steam generator.

**O**n October 31, 1957, two of Frisco's racehorse E units, 2005 - *Winchester* and 2003 - *Steel Dust*, have been serviced and await the call for their next run. Among other details to consider here are the flatcar load to the right and the steam deadline forming behind the shop.

**E**ven passenger units need sand to start a train. The 2003, named *Steel Dust*, was originally an E7 built with fluted stainless-steel side panels. When built, the 2003 was assigned to the METEOR and TEXAS SPECIAL streamliners. After rebuilding with standard E8 side panels, the characteristic E7 louver behind the cab door was retained.

# Condensed Freight

## BETWEEN ST. LOUIS AND...

Also photographed on October 31, 1957, lone GP7 545 was parked alongside the round house.

A deck girder bridge over Lansdowne Street at the St. Louis city limits is an appropriate location for advertising passenger service.

# Schedules

**P**oking out from beside the diesel shop on February 13, 1958 was FA1 5220 and F3A 5010. The 5010 was rebuilt in kind by General Motors to an F7 and classified by the Frisco as an F37A.

**P**arked in the shade on a hot August 29, 1954 was lead unit 5219, an FA1. In three years, many of the FA/FB fleet, including the 5219, would be re-engined by General Motors with 16-567C diesel engines, essentially creating F9m's. Working the outside lead of the freight yard, NW2 259 approaches the camera.

South of St. Louis on the Memphis Division, line changes often isolated facilities no longer deemed necessary to operations. One such location was Meramec Highlands. On January 25, 1955 the depot, bypassed by a 1933 line change still stood in haunting solitude. After the bypass was finished, the depot was used for few a years as a residence. It and a nearby tunnel were often visited by Jim on his photo excursions. The tunnel was even used as a mushroom farm for a while in the early 1950s.

At mile post 12.5 south of St. Louis, on a very sunny November 7, 1955, southbound train No. 807 heads for Memphis. THE SUNNYLAND was pulled by E8 2020, appropriately named *Big Red*.

FRISCO
RAILWAY EXPRESS AGENCY

FRISCO

# ST. LOUIS SOUTHWESTERN

The Cotton Belt, a Southern Pacific subsidiary, almost went out of its way to be anonymous in the St. Louis area. Operating over Missouri Pacific trackage from the Mississippi crossing at Thebes, Illinois, it came into town over the Mopac's Chester Subdivision to Valley Junction, north of Dupo. At Valley Junction, Cotton Belt freights headed east to Valley Yard, their yard and engine terminal, nestled between Illinois Central's "B" yard and Alton & Southern's Davis Yard. Passenger trains went west at Valley Junction to Union Station. Passenger units, two PA1s, a GP7 and an FP7, were serviced by the TRRA at Fourteenth Street.

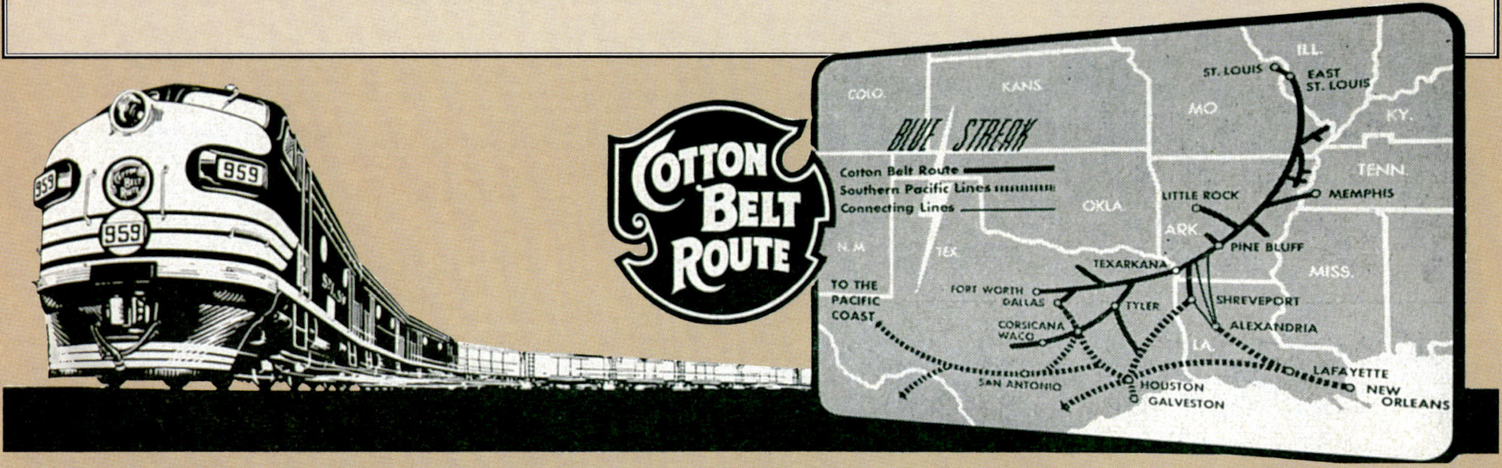

On May 19, 1956 Cotton Belt FP7 306 waits near the throat of the wye on the east side of Union Station for permission to back to her waiting train. The DAYLIGHT paint scheme is a sure give-away as to the affiliation of Cotton Belt and Southern Pacific.

# Terminal Railroad Association of St. Louis

The TRRA of St. Louis, a terminal transfer and switching company is a co-operative company equally owned by the following trunk lines:

- •Baltimore & Ohio - Southwestern
- •Chicago, Burlington & Quincy
- •Chicago, Rock Island & Pacific
- •Cleveland, Cincinnati, Chicago & St. Louis *(Big Four) part of the New York Central*
- •Gulf, Mobile & Ohio *(Chicago & Alton was member prior to merger)*
- •Illinois Central
- •Louisville & Nashville
- •Missouri-Kansas-Texas
- •Missouri Pacific
- •Pennsylvania
- •St. Louis-San Francisco *(Frisco)*
- •St. Louis Southwestern *(Cotton Belt)*
- •Southern
- •Wabash

The TRRA operates facilities on both sides of the Mississippi River, connected by the Eads and Merchants Bridges. TRRA is the only company that has connections to all the railroads on both sides of the River. Though the company is made up of numerous subsidiaries, they are arranged into three operating companies.

Terminal Railroad owns and operates Union Station and, in co-operation with Merchants Bridge Terminal, provides passenger train routes to that Station. In conjunction with the operation of Union Station, yards and engine facilities to service passenger equipment are provided. It handles freight interchange in the Mill Creek Valley and in East St. Louis via Eads Bridge using a large clearing yard north of Relay Junction, known as C. D. Yard. All connections on the west side of the Mississippi deliver to Twenty Third Street Yard in Mill Creek Valley.

Merchants Bridge Terminal Railway operates the Illinois Transfer Railway (East St. Louis Outer Belt), which extends around East St. Louis from Merchants Bridge to Valley Junction on the south. Merchants Bridge lines operated into Granite City and Madison, where a large classification yard was located. On the west bank of the Mississippi, it operated the River Front High Line to the Merchants Bridge, several smaller yards along the river on the north side of town and the Outer Belt line on the north and west side of town, connecting to the Terminal Railroad facilities at Grand Avenue. Merchants crews deliver to Mopac's Dupo Yard via Valley Junction. All other traffic east of the Mississippi is handled out of Madison Yard, north of Merchants Bridge.

Wiggins Ferry Company operates the river front lines and industrial trackage on both sides of the Mississippi River. Originally, connections between the two sides of the river were made with car ferries. This was eventually superseded by the all-rail line routing via the Eads Bridge from East St. Louis and through Madison, Illinois via the Merchants Bridge. The main clearing yard on the east bank of the river connected to all the trunk lines that terminate in East St. Louis. On the west side, Wiggins Ferry connects with the Missouri Pacific, Manufacturers Railway, CB&Q and the Wabash.

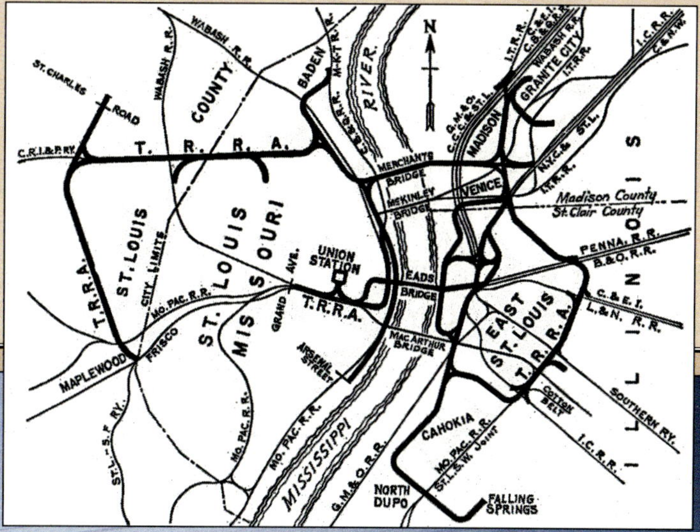

**O**n May 26, 1956 TRRA 1603 pulls a transfer past Relay Junction, headed for the classification yard just to the north, C. D. Yard. Orphans among the switcher fleet, the four RS3 road switchers, numbered 1600-1603, were always kept on the Illinois side working interyard transfers.

**W**iggins Ferry trackage on the west side of the river ran along the top of the stone-faced levee which protected the waterfront. The bridge to the south is the MacArthur Bridge, owned by the city. Looking north past that American Refrigerator Transit mechanical reefer is the Eads Bridge, owned and operated by the TRRA. These vintage cars were recorded on December 1, 1957. This line interchanged with the MP Lesperance Yard south of the MacArthur Bridge, which explains the high percentage of MP cars in this group.

Northbound on the Merchants Bridge High Line is this cab hop consisting of SW1200 1210 and transfer caboose 609. This scene was recorded on September 4, 1960.

Southbound along the High Line on September 13, 1956 was this company "passenger train" consisting of an SW1 switcher and truss-rod equipped coach 25. The coach was used to ferry employees from the Brooklyn Yard and shops in Illinois to the Eighteenth Street coach yard and the Fourteenth Street engine terminal. This view illustrates the River Front High Line and two level (railroad on lower deck) MacArthur Bridge at their best.

NICKEL PLATE ROAD
181

**A** superb example of the diversity of passenger power seen at the TRRA Fourteenth Street engine terminal was recorded on December 22, 1955. From the left, power waiting and being serviced include Nickel Plate PA1 181, Cotton Belt FP7 306, a MKT PA1, a New York Central J1 Hudson and several Terminal Railroad switchers. In the background are fuel oil tank cars from the refineries at Wood River, Illinois, just south of Alton on the CCC&StL. Fourteenth Street went through six or eight cars of diesel fuel daily during the fifties.

**T**he view to the east of the Eighteenth Street Viaduct on October 6, 1957 found the wooden transfer coach, No. 25, parked on the outside track of the head end yard. The mail and express terminal was just to the east of Union Station and this yard provided storage for cars being held for loading.

**B**aldwin VO-1000 531 was photographed from the Fourteenth Street Viaduct on September 2, 1954. The string of MP baggage and mail cars have just been pulled from the mail terminal on the east side of Union Station.

**A** time exposure taken in the train shed of Union Station on March 8, 1958. Reflecting the drumhead for the GM&O ABRAHAM LINCOLN to Chicago, Missouri Pacific sleeping cars are parked awaiting early patron arrivals.

**O**n the evening of May 21, 1958 VO 1000 591 works Terminal Railroad trackage at the eastern throat of Union Station. A trio of Pennsy E-units wait near the sand towers below the viaduct.

**L**ooking east from the Jefferson Avenue Viaduct on May 30, 1957 EMD SW1 504 was spotted between moves. In the background, a Harriman-style Illinois Central consist is nearest the camera.

## USE THE ST. LOUIS GATEWAY

**NO CHANGE OF STATIONS - - NO STEPS TO CLIMB**

On May 30, 1957 passenger cars being switched on Terminal Railroad trackage west of Jefferson Avenue represent a national cross section of trains and equipment. The Union Pacific equipment was for the joint UP/Wabash CITY OF ST. LOUIS. Older Pullmans, represented by the Northern Pacific's *Chief Rocky Bear*, were pooled for special movements such as for the Boy Scouts or Shriners conventions. Missouri Pacific tracks were parallel to the Terminal Railroad yards west of Twenty third Street, which explains the Mopac Business Cars alongside the Pullmans.

**A**pproaching from the west, Baldwin VO 1000 533 pulls a cut of cars between the Rankin Avenue coach yard and Jefferson Avenue. As Jim grew up prior to World War II, many families, including his parents would spend weekend afternoons watching the yard switching operations from 21st Street Viaduct. These family outings were the sort of activity that fueled Jim's interest and caused him to return years later with a camera.

**C**aught slowly shoving a carload of replacement wheel sets, Baldwin VO 1000 598 passes the car washing facility at Ewing Avenue on September 24, 1956. Built in October 1944, the 598 would last until 1965 when it would be traded for a new EMD SW1200.

Looking east along the yard lead, the Jefferson Avenue coach yard has two SW1 switchers busy moving cars on the afternoon of September 24, 1956. The view is from the Ewing Avenue Viaduct.

West of Ewing Avenue, the 598 is passed SW1 505 and a string of head end equipment on September 24, 1956. The boxcar next to the 505 is a converted U. S. Army Troop Sleeper in express service for the New York Central.

# WABASH RAILROAD

Wabash served St. Louis over two routes, the Decatur Division from the northeast, and the Moberly Division from the west. This combination gives the Wabash an east-west through line via trackage rights over the Merchants Bridge, the only such line in the region. The Moberly Division, St. Louis District, saw freight and passenger traffic from Omaha, Des Moines and Kansas City. Decatur Division, Brooklyn District, trackage starts at Decatur, Illinois, the junction for lines from Chicago and Detroit funneling to the southwest.

The Decatur Division trackage passes through Edwardsville and crosses the Illinois Terminal at Bluff Junction. South of Bluff Junction the line crossed Mississippi bottoms land, running southwest to Mitchell where it parallels the GM&O/Big Four joint trackage to Granite City. At WR Tower the lines split, freight going south over the TRRA to Brooklyn Yard while passenger trains cross over the Merchants Bridge to St. Louis. West of the Mississippi the line swings north to Luther Yard. Passenger traffic bypassed Luther Yard and went west over the TRRA Merchants Belt line to Easton Junction. There trains head south over the Wabash (Moberly Division) line through Delmar Station to Grand Avenue Tower. East of Grand Avenue, traffic used the TRRA Terminal line to Union Station. Trackage rights between Easton Junction in north St. Louis and WR Tower in Granite City completed the through line. Westbound freight was dispatched out of Luther Yard over the TRRA to Easton Junction. Passenger power was serviced at the Wabash Engine Terminal at Vandeventer and Clayton Avenues in St. Louis.

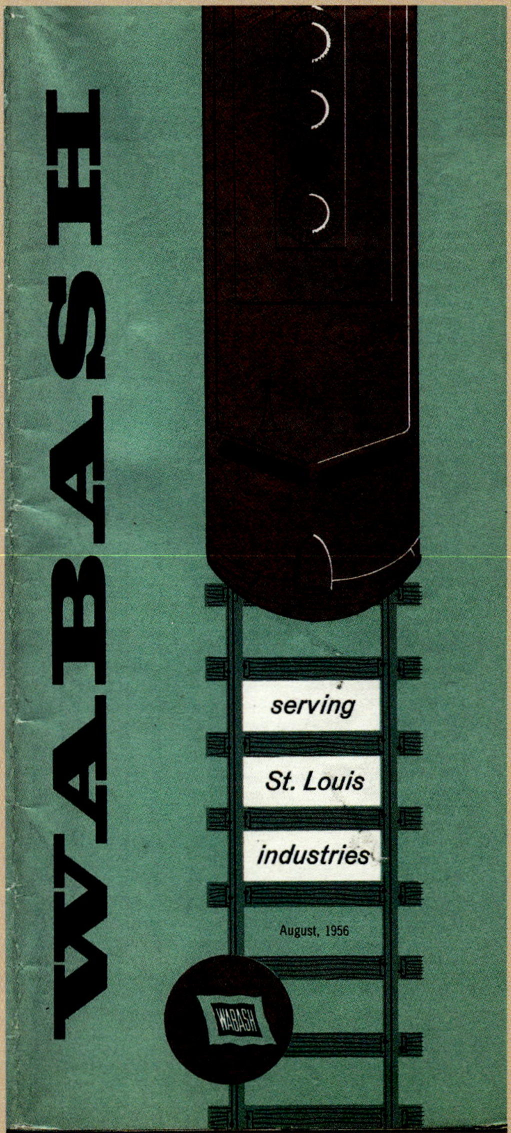

**A** rear brakeman enjoys a brief reflective moment during a station stop. His assigned post was the rear platform on the Observation/Parlor Car *City of Wabash*, the rear car of THE BANNER BLUE on October 1, 1959.

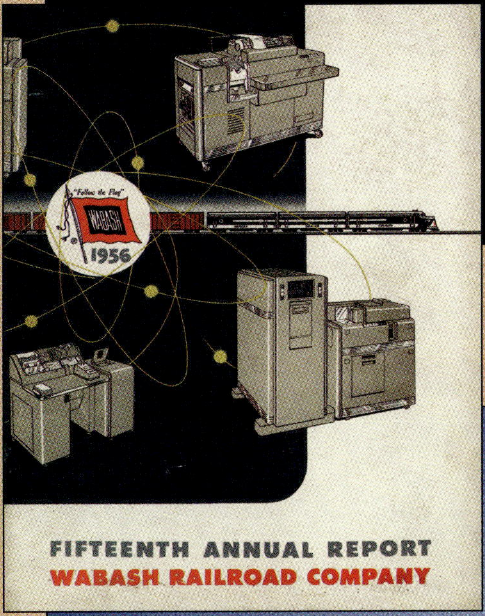


**E**astbound at Mitchell on January 3, 1956, THE CANNONBALL is powered by EMD E8 1006 and 1004. This track belongs to the double tracked Decatur Division.

**B**ringing up the markers for the eastbound CANNONBALL, Observation/Parlor Car *City of Lafayette* rolls north through Mitchell on January 3, 1956.

From a hill overlooking Lenox Junction, the layout of the junction is displayed before us. The Wabash train, THE BLUE BIRD, is on the eastbound track of the double tracked Decatur Division. Sentinels from the past, the bridge piers supported the Illinois Terminal interurban line from Granite City to Alton until the line was removed in 1953. Beyond the piers is Lenox Tower where the CCC&StL lines split, the original line running parallel to the GM&O north to Alton, the "cutoff" turning east to Hillsboro. The distant industrial complex includes the Shell and Standard Oil refineries at Wood River, Illinois.

**A** westbound freight lead by F7 1169A rolls through Mitchell on January 3, 1956. This train will tie up at the Brooklyn Yard after heading south on the TRRA at WR Tower on the south side of Granite City.

The main passenger engine terminal for the Wabash was near the intersection of Vandeventer and Clayton Avenue. On April 17, 1956, two sets of E7s wait near the sand tower, lead by the 1002 and the 1001A respectively.

**W**abash E8 1015 was parked on the turntable lead on March 31, 1958. This was a largely industrial part of town as can be seen from the background. By this time, Jim was nearing the end of his college experience and had the use of his parent's automobile so this industrial neighborhood did not pose the problems of access that earlier photo excursions might have encountered.

1002

Framed by two units, the Vandeventer passenger depot was a busy spot on a gray February 21, 1956. To the left is E7 1002 waiting for the passage of westbound E8 1003 near the depot.

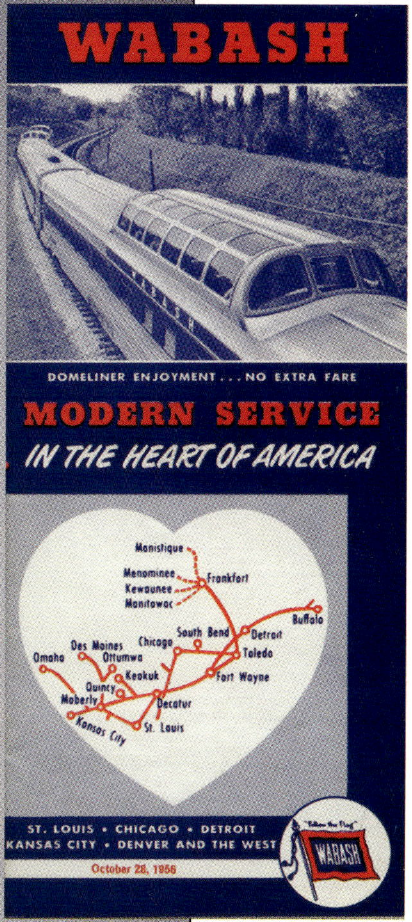

**F**acing east, Alco-GE PA1 1020 is parked at the sanding tower for service on January 13, 1955. Snow did not normally stay on the ground long, since St. Louis is pretty far south. This was one of those cold days when it did. EMD switcher 101 is just west of the supports for the still standing coaling tower.

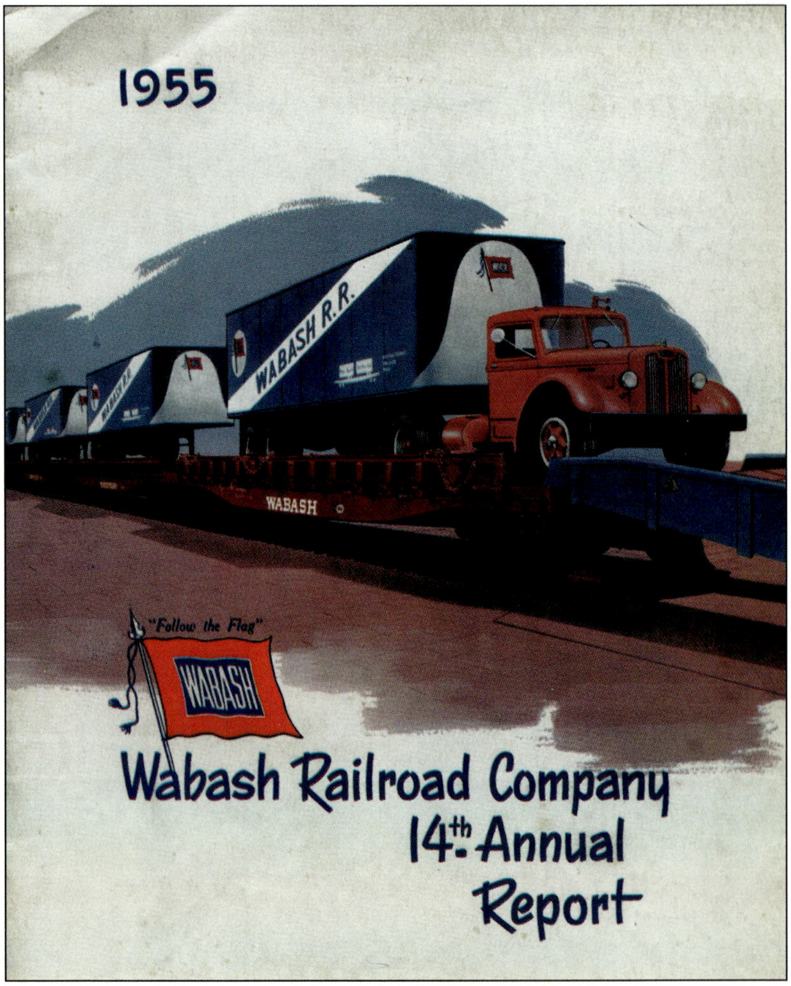

In the same location as the 1015, PA1 1021 is parked west of the Clayton Avenue roundhouse on March 28, 1956. Looking east are the support frame for the power leads to the turntable and the sand tower serving tracks coming off the east end of the turntable.

1015
1015

The concrete coaling tower was still standing on January 13, 1955. Parked on the leads adjacent the tower are E8 1015 and PA1 1020. In some of the other views taken the following year the tower has already been removed.

**A** caboose-hop on the Wabash was photographed on September 4, 1960.
Westbound at Grand Avenue Tower GP7 478 leads caboose 2603 through the interlocking.

**D**elmar Station as seen from the south on April 30, 1957. The train in the depot is the eastbound BLUE BIRD.

**T**he neon signs atop Delmar Station were photographed on the night of March 3, 1956. This station is on the Moberly Division trackage in St. Louis, west of Grand Avenue Tower.

551
WABASH

10
Banner
Route

**A** rarity at Union Station, two Fairbanks Morse "Train Master" units, the 550 and 551, back into the trainshed on September 14, 1956. The "Train Masters" are on the west leg of the wye and TRRA 1208 is are on the east leg. Multi-tracked, interlaced wyes allowed trains to access any of the 32 station tracks within Union Station's train shed.

**I**n an unusual application of the "Banner" slogan, a baggage car on one of the Kansas City trains sports a huge BANNER ROUTE as it is shoved back around the wye at Eighteenth Street. On September 14, 1956 this rolling billboard was the first car out behind the 1020A and 1006 as the train was turned prior to departure.

Having its start in the winter of 1946, this collection came to be at Barretts Station, Missouri through the cooperation of the Missouri Pacific. In the early years, local roads made up the core of the collection with donations of recently retired equipment. Jim discovered the Museum in November 1949 while on a Sunday drive with his parents following a tip from Lewis F. Carna Jr., son of his employer at the Carna Photo Studio. The Museum is where Ozment and his contemporaries learned much of their railroad history and honed their future interests. Jim stopped visiting the Museum as a volunteer in 1955, preferring to spend more time as a photographer of the rest of the St. Louis rail scene. Many of the early volunteers still spend their time helping out at the Museum, now administered by the St. Louis County Parks and Recreation Board. We have included an early sampling of what developed into one of the nation's best known transportation museums.

Caught in the anxious pose of a hoghead waiting for a departure highball, Jim Ozment was at the throttle of Wabash 573 on August 18, 1955. Part of the early collection, the F-4 2-6-0 served the young museum volunteers as a full size training simulator to learn the operations of steam power.

Close friend and lifelong volunteer at the Museum of Transport, Willis R. Goldschmitt poses with an ancient Mack Truck on the museum grounds. Jim recorded this advertising simulation on November 26, 1954.

**A** recent arrival, Illinois Terminal Railbus 206 was photographed on November 4, 1954. Parked along with PSC 615 and IT 241, Jim rode 206 prior to its retirement only a few months before. Self-propelled railbuses replaced steam-powered trains on the Alton to Grafton (Illinois) line in 1933. The 206 came to the Museum of Transport after service west of Alton was terminated on March 7, 1953.

**W**earing a green and red scheme that preceded the standard "Traction Orange" of her then in-service sisters, St. Louis Water Works Railway 17 was at Barretts on September 17, 1953. As reported earlier in this volume, the remainder of the fleet worked under wire until April 1955.

**O**ne of the better pieces in the early collection, Rock Island caboose 18058 was photographed on September 17, 1953. This car served as a bunk car for the young volunteers who would spend weekends working on equipment or the museum grounds.

**A** notable prize for the Museum of Transport came as a personal donation from John W. Barriger III. While president of the Monon in the 1940s, Barriger used this heavyweight business car as a traveling office on inspection trips. Upon his retirement, Monon Business Car 90, shown here on September 17, 1953, was donated to the Museum of Transport. The car then served as the longtime residence of Dr. John and Mary Roberts, founders of the Museum.

**A** group of friends gathered inside Rock Island 18058 for this portrait on November 25, 1954. This generation of railfans started out as volunteers at the Museum of Transport. From the left this group includes Jim Ozment, Carl Horn, Warren Hertel, Artie Brown and Mel Olsen.

**O**ne of the last schools in the nation to offer a degree in Railway Engineering, Purdue University also maintained a collection of rolling stock. That collection was transferred to the Museum of Transport in 1953. One piece from the Purdue Collection was this trolley-equipped test car, formerly used on the numerous electric interurban systems in Indiana, and shown here in September 1953.

An overall view of the Museum grounds looking north was taken from atop West Barretts Hill on September 17, 1953. The maintenance of way piece in the foreground was a 5-foot gauge steam shovel from the St. Louis, Iron Mountain & Southern, a Missouri Pacific predecessor.

A corporate sponsor was a rare commodity in the 1950s. Most museum operations were at the mercy of individual donors and part-time volunteers, so it is understandable that many of them were dissolved and their collections scattered. This sign announcing the assistance of the Texas Oil Company and the use of their lubricants gave an indication of the longevity the Museum of Transport would enjoy. Jim photographed the sign on February 28, 1955.

This view of the Museum is toward the west and was taken from the south side of East Barretts Hill on December 31, 1954. At the far left of the scene is West Barretts Tunnel, bypassed with the 1940s line change, now part of the Museum of Transport grounds.

# Into the Sunset

**S**outh of Dupo on the Illinois side of the Mississippi, Missouri Pacific Chester Subdivision trackage hugs the river as it heads for the crossing at Thebes, Illinois. Jointly operated with the Cotton Belt, this is fast, heavy duty mainline. On September 4, 1954, Jim caught the reflected sunset off the river and the Mopac south of Chester, Illinois, for which the subdivision was named.